P9-AQS-710

The Pluto Files

Also by Neil deGrasse Tyson

Death by Black Hole and Other Cosmic Quandaries
(2007)

The Sky Is Not the Limit:
Adventures of an Urban Astrophysicist
(2004)

Origins: Fourteen Billion Years of Cosmic Evolution
(with Donald Goldsmith)
(2004)

My Favorite Universe
(a 12-part video lecture series)
(2003)

Cosmic Horizons: Astronomy at the Cutting Edge
(with Steven Soter, eds.)
(2001)

One Universe: At Home in the Cosmos
(with Charles Liu and Robert Irion)
(2000)

Just Visiting This Planet
(1998)

Merlin's Tour of the Universe
(1997)

Universe Down to Earth
(1994)

Neil deGrasse Tyson

W. W. NORTON & COMPANY
NEW YORK LONDON

THE PLUTO FILES

The Rise and Fall of America's Favorite Planet

Printed in the United States of America
First Edition

For information about special discounts for bulk purchases, please contact W. W. Norton Special Sales at specialsales@wwnorton.com or 800-233-4830

Manufacturing by R R Donnelley, Crawfordsville
Book design by Lovedog Studio
Production manager: Julia Druskin

Library of Congress Cataloging-in-Publication Data

Tyson, Neil deGrasse.
The Pluto files : the rise and fall of America's favorite planet / Neil deGrasse Tyson.
p. cm.
Includes bibliographical references and index.
ISBN 978-0-393-06520-6 (hardcover)
1. Pluto (Dwarf planet) I. Title.
QB701.T97 2009
523.49'22—dc22
2008040436

W. W. Norton & Company, Inc.
500 Fifth Avenue, New York, N.Y. 10110
www.wwnorton.com

W. W. Norton & Company Ltd.
Castle House, 75/76 Wells Street, London W1T 3QT

1 2 3 4 5 6 7 8 9 0

To Plutophiles young and old

3-26-08

Dear Dr. neil Tyson degrasse
At first, remember all of those kids that send you bad letters? Well, I want to apolijize all the things that we were wrong about.
We're sorry about...
giving you mean letters.
saying we love pluto but not you.
I'm very sorry, it'll be ok.

Taylor
age 7

PLUTO

Letter from Taylor Williams, Mrs. Koch's second-grade class, Roland Lewis Elementary School, Tampa, Florida (spring 2008).

Contents

WHOSE TERRITORY ARE THE EARTHLINGS FEUDING OVER NOW??
OURS.
PLUTO
WWW.LOHUD.COM/BLOGS/DAVIES
© 2006 THE JOURNAL NEWS
MATT DAVIES

Preface

GATHERED HERE IN ONE PLACE IS A RECORD OF PLUTO'S RISE AND fall from planethood, given by way of media accounts, public forums, cartoons, and letters I received from disgruntled schoolchildren, their teachers, strongly opinionated adults, and colleagues.

In February 2000, the American Museum of Natural History opened its $230 million Frederick Phineas and Sandra Priest Rose Center for Earth and Space, containing the rebuilt Hayden Planetarium, on the corner of 81st Street and Central Park West in New York City. The newly conceived exhibits treated the solar system in a way that was without precedent for public institutions, even though murmurs had already begun in the planetary science community that something needed to be done about Pluto's classification in the solar system.

The exhibit models, their accompanying text, and the overall layout of the Rose Center organized the principal contents of the solar system by objects of like properties, rather than as enumerations of planets and their moons. This decision landed Pluto among the growing number of icy objects found beyond Neptune and left it unmentioned and out of view among our models for the rocky, terrestrial objects (Mercury, Venus, Earth, and Mars) and the gas giants (Jupiter, Saturn, Uranus, and

Neptune). By this organization, we practically abandoned the concept of planet altogether.

This decision represented the consensus of the science committee for the Rose Center's design and construction, of which I served as head. While the accountability and originality of our pedagogical approach to the subject lies equally among us on the committee, as director of the Hayden Planetarium I became the most visible exponent of this decision when, a full year after the Rose Center opened to the public, the *New York Times* broke a page 1 news story that we had "demoted" Pluto from its ranks of planethood. I was thenceforth branded a public enemy of Pluto lovers the world over.

This distinction prevailed until August 2006, when the International Astronomical Union (IAU), prompted by pressure from the professional community of planetary scientists as well as from the general public, brought Pluto's planethood to a vote at a triennial general assembly in Prague, Czech Republic. The result? Pluto was formally downgraded from "planet" to "dwarf planet," thereby helping to diffuse the negative attention that we had been receiving for six years running.

It's one thing for a single institution to reexamine Pluto's standing in the solar system, but it's quite another for an international organization of astronomers to do so. When the IAU voting results were released, a media frenzy followed, temporarily displacing news stories on terrorism, the Iraq War, genocide in Darfur, and global warming.

The Pluto Files chronicles and documents Pluto's remarkable grip on the hearts and minds of the American public, the professionals, and the press.

Neil deGrasse Tyson
New York City
October 2008

The Pluto Files

SHENEMAN The Star-Ledger
PLUTO TIMES
EARTHLINGS DECLARE PLUTO NOT A PLANET
"I DON'T SEE HOW ANY PLANET THAT ALLOWS PARIS HILTON TO RECORD A CD HAS ANY RIGHT PASSING JUDGMENT ON US."

1

Pluto in Culture

At about four in the afternoon on February 18, 1930, 24-year-old Clyde W. Tombaugh, a farm boy and amateur astronomer from Illinois, discovered on the sky what would shortly be named for the Roman god of the underworld. Tombaugh had been hired by Arizona's Lowell Observatory to search for the mysterious and distant Planet X. The observatory was named and founded in 1894 by Percival Lowell, an independently wealthy American astronomer, who died in 1916, but not before launching the search that Tombaugh would complete. On March 13, 1930, Lowell Observatory went public with the news.

Shortly thereafter, two well-known architectural icons were out-of-date the day they were dedicated. On May 12, 1930, a mere two months after Pluto's discovery, the Adler Planetarium, on Chicago's South Lake Shore Drive, opened for business—the first of its kind in the Western Hemisphere[1] and today the oldest surviving planetarium in the world. Adler's ornate entrance lobby had been designed well before Pluto's discovery and displays a circle of plaques affixed to the wall that duly identifies a Plutoless family of eight planets in the solar system.

And in New York City, between Fiftieth and Fifty-first streets along Fifth Avenue, you will find, across the avenue from the main entrance to Saint Patrick's Cathedral, a large and mighty brass statue of Atlas, designed by the sculptor Lee Lawrie in the 1920s and erected in the 1930s as part of the extensive art deco Rockefeller Center complex. You may remember from mythology class that for his misdeeds, Atlas was condemned by Zeus to stand on Earth's western edge and hold the entire sky on his shoulders, preventing Earth and sky from resuming their primordial embrace. To represent the sky, Lawrie molded a spherical celestial grid, as is common for such artistic needs. But across Atlas's yoke, Lawrie identified the symbol for each of the planets, Earth's Moon included, lest you were not yet convinced that the known universe is what Atlas carried. Of course, in the 1920s Pluto had not yet been discovered, so once again Pluto was too late for the party. Atlas's yoke identifies Mercury through Neptune, with no room for a ninth planet. No room for Pluto.

The world of music would be similarly afflicted.

In search of cosmic themes for his next orchestral work, the English composer Gustav Holst (1874–1934) wrote his seven-movement masterpiece *The Planets* in 1916. Holst drew his musical themes from the lives

1. New York City's Hayden Planetarium, dedicated in October 1935, followed that of the Buhl Planetarium in Pittsburgh and the Griffith Observatory and Planetarium in Los Angeles.

Figure 1.1. *Adler Planetarium, Chicago. It was dedicated only two months after the announcement of Pluto's discovery.*

Figure 1.2. *The sunlit, spectrum-graced entrance foyer to the Adler Planetarium, in Chicago. Planetarium CEO Paul Knappenberger (left) and the author flank the eight bas-relief plaques from 1930, one for each planet of the solar system—except Pluto, which had not been discovered at the time the plaques were designed and cast.*

Figure 1.3. *The statue of Atlas stands tall and mighty, in art deco splendor, along Fifth Avenue at Rockefeller Center, in New York City. The solar system labeled across his shoulders is missing Pluto. The statue was designed by Lee Lawrie in the 1920s, before Pluto was discovered.*

Figure 1.4. *Atlas, detail. Rising above Atlas's six-pack abs and his bulging biceps we see the yoke that displays in relief the eight planets of the solar system, plus the Moon. From right to left we have the symbols for Mercury, Venus, Moon + Earth, Mars, Jupiter (eclipsed by Atlas's linebacker neck), Saturn, Uranus, and finally Neptune.*

and times of the Roman mythological characters after whom the planets were named. Of course, the music is absent a movement dedicated to Pluto, which had not yet been discovered, and it's missing Earth, which was not a classical planet, leaving a count of only seven.

Shortly after Clyde Tombaugh's discovery, Holst began working on a Pluto movement, inspired, of course, by themes of the underworld. He partially completed the work before suffering a stroke. But after an effort to dictate the rest of the movement to one of his students, Holst (with unwitting foresight for Pluto's fate) abandoned the effort, unhappy with how the work was turning out.

Not letting the music of the spheres rest, composer and Holst scholar Colin Matthews wrote the "missing" Pluto movement in 2000 for the Manchester-based Hallé Orchestra. But with Pluto's 2006 demotion to the status of "dwarf planet," Matthews's notes, though well meaning, might better serve as the first movement of a yet-to-be-written orchestral work that celebrates the exotic icy bodies of the outer solar system.

While Clyde Tombaugh was searching for Planet X, the roaring twenties was in full swing. At the time, most Americans associated the name Pluto with the commercial product Pluto Water, a heavily advertised and widely used mineral water laxative. Promising "Relief for Constipation in 30 minutes to 2 hours," Pluto Water was bottled on the grounds of the palatial French Lick Springs Hotel in Indiana, about 50 miles south of Bloomington. In a hard-to-forget slogan, the Pluto Water ads further proclaimed, "When Nature Won't—Pluto Will." So you wouldn't have expected Americans of the day to come up with the name Pluto for the newly discovered cosmic object. And they didn't.

Percival Lowell's widow, Constance, proposed the name Percival for the new object, which would sound odd to many astronomers' ears. Yet this gesture was not the first expression of cosmic audacity in the

Figure 1.5. A 1932 advertisement for Pluto Water, a popular laxative in America around the time Pluto the planet was named by an 11-year-old girl from England.

history of planet names. After the English astronomer William Herschel (1738–1822), who we will learn more about later, was convinced he had discovered a real planet in 1781 (the first ever discovered by anyone), he did what any good citizen of an aristocracy would do: he named his new planet after King George III. For years, the planets of the solar system would be identified as Mercury, Venus, Earth, Mars, Jupiter, Saturn, and Georgium Sidus. I don't know about you, but I find something unsettling about a planet named George, even if he is a king. Apparently, so did everybody else. Eventually, the Roman nomenclature was resumed, leading to the name Uranus, Roman god of the sky, and son and husband of mother Earth.

By tradition, dating back to Galileo in the early 1600s, planet moons are named for Greek mythological characters in the life of the Greek god whose Roman counterpart is the name of the planet itself. For example, the four brightest moons of Jupiter—Io, Ganymede, Callisto, and Europa—are characters in the life of the Greek god Zeus, for whom Jupiter is the Roman counterpart. But in the lone exception to this rule, partly appeasing the British people, who were dissed by not having a planet they discovered named for their king, the moons of Uranus are named for characters in Shakespearean plays. Among them we find Ariel, Caliban, and Miranda (all from *The Tempest*), Oberon and Puck (both from *A Midsummer Night's Dream*), and Bianca (from *The Taming of the Shrew*).

The name Pluto was first suggested over breakfast on Friday, March 14, 1930, by Venetia Burney, an 11-year-old schoolgirl in Oxford, England, after her grandfather had read the news story that Lowell Observatory discovered a new planet. Unlike Americans across the Atlantic, Venetia probably never used or even heard of Indiana's Pluto Water laxative, leaving her free from scatological bias against the name. She had been studying classical mythology in school and of course knew the other planets. With the name Pluto not yet taken, she blurted out to her grandfather, "Why not call it Pluto?,"[2] knowing that Pluto is, after all, the god of the dead and underworld, the realm of darkness. And what else, if not darkness, prevails 4 billion miles from the Sun?

The rest is history. Or rather, good luck. Venetia's grandfather, Falconer Madan, was a retired librarian from the Bodleian Library of Oxford University who happened to be friends with many astronomers. Madan suggested the name to Herbert Hall Turner, Oxford professor and former astronomer royal (who, among other credits, coined the term *parsec*),[3]

2. NASA interview, January 2006; www.NASA.gov/multimedia/podcasting.

3. A unit of distance in astronomy equal to 3.26 light-years, itself equal to about 19 trillion miles and derived from the distance a star would have to be for it to exhibit a parallax angle of 1 second of arc (hence par-sec) against the background stars as Earth orbits from one side of the Sun to the other.

Figure 1.6. Venetia Burney, an 11-year-old schoolgirl in Oxford, England, first suggested the name Pluto after her well-connected grandfather had read the news story of Lowell Observatory's discovery of a planet. She had been studying classical mythology, and Pluto, the god of the underworld, was fresh in her mind.

and Turner promptly cabled the name to fellow astronomers at the Lowell Observatory.

Other suggestions for names included Artemis, Atlas, Constance, Lowell, Minerva, Zeus, and Zymal. But Pluto eventually triumphed. The name also maintains a happy family with Jupiter and Neptune as Pluto's brothers in Roman mythology.

Apparently, the naming of cosmic objects was already in Venetia's bloodline. Her great-uncle, Henry Madan, was science master of Eaton College in 1877 when he named the two newly discovered moons of Mars—Phobos (Fear) and Deimos (Terror)—after the battle companions to Ares, the Greek god of war. Venetia Burney ultimately became Venetia

Burney Phair and worked as an economics teacher before retiring to her home in Epsom, England.

Lowell Observatory officially proposed the name Pluto on May 1, 1930, in simultaneous letters to the American Astronomical Society, the Royal Astronomical Society, and the *New York Times*. Pluto's official symbol would be the overlapping, juxtaposed letters P and L, the first two letters of Pluto and, in a happy coincidence, the initials of Percival Lowell, who instigated the search in the first place.

Eleven years later, in 1941, a team of physicists led by Glen T. Seaborg manufactured a new element for the Periodic Table (you may remember this ubiquitous grid of boxes from chemistry class) while working at the

Figure 1.7. *Herbert Hall Turner, Oxford professor and former astronomer royal, promptly cabled the name Pluto to fellow astronomers across the Atlantic at the Lowell Observatory after hearing the suggestion from 11-year-old Venetia Burney, via her grandfather, Falconer Madan, a retired librarian from the Bodleian Library of Oxford University.*

University of California at Berkeley's cyclotron—one of the world's preeminent atom smashers. The new element had 94 protons in its nucleus and was in need of a name. There, in the outer solar system, the newly discovered planet loomed large. Thus was born plutonium. This fissionable element became the active ingredient in the atomic bomb that the U.S. Army Air Force dropped over the Japanese city of Nagasaki on August 9, 1945, just weeks after the bomb had been tested on July 16, 1945, at the Trinity test site in New Mexico—the first-ever detonation of a nuclear weapon. The one dropped over Hiroshima on August 6 was not pretested. It used uranium, whose bomb-worthy fissionable properties had been well established on paper and in the laboratory.

In hindsight, plutonium was destined to be named for Pluto. By 1789, just eight years after Herschel discovered Uranus, Martin Klaproth, of Germany, discovered the heaviest atom in nature. In need of a name, and with the planet Uranus fresh on people's minds, the element uranium would ultimately land at slot 92 on the periodic table, harboring 92 protons in its nucleus.

Before you even discovered the next element in sequence, what would you want to name it? Berkeley physicists Edwin M. McMillan and Philip H. Abelson discovered element number 93 in 1940 and duly named it neptunium after planet Neptune, thus mirroring the sequence of element names to the run of planet names in the outer solar system, leaving plutonium to follow naturally thereafter.

In spite of its later-to-be-determined diminutive size, Pluto, the god of death, is forever enshrined on our periodic table of elements and associated, by name, with the atomic bomb, one of the greatest weapons of destruction ever devised.

The Periodic Table has memorialized other cosmic objects as well. The first two asteroids discovered, Ceres and Pallas, led to cerium and palladium. Earth and Moon are there, too, in the guise of the rare elements tellurium and selenium (from the Latin *Tellus* for Earth and the Greek *Selene* for Moon), found naturally together in ores.

Meanwhile, back in Los Angeles on September 5, 1930, the fledgling Disney Brothers Studio releases a cartoon titled "The Chain Gang," featuring two bloodhounds hot on the trail of Mickey Mouse, an escaped convict. These unnamed canines would serve as the model for the character who would become Pluto, Mickey's pet dog, but not before some further experiments with the concept.

On October 23, 1930, Disney releases "The Picnic," featuring a bloodhound character, but with the name Rover, who in this cartoon belongs to Minnie Mouse. Both Rover and Minnie join ex-convict Mickey for a picnic. Minnie wants to eat. Rover wants to play. And Mickey, having spent so much time in jail, is horny. But Rover keeps preventing amorous encounters between Mickey and Minnie, angering Mickey. Rover makes amends by using his tail as a windshield wiper when Mickey and Minnie drive home in a rainstorm.

At last, on May 3, 1931, Disney releases "The Mouse Hunt," in which the playful bloodhound first appears as Pluto, Mickey's dog. In a press release issued by Mickey Mouse, the rodent recalls Walt Disney suggesting the alliterative Pluto the Pup:

> *Walt decided that I should have a pet and we decided on a dog. All the writers at Disney tried to come up with a name. We tried the "Rovers" and the "Pals", but none seemed to fit. Then one day, Walt came by and said, how about Pluto the Pup? And that's what it's been ever since.*[4]

After twenty or so cartoon appearances, Pluto finally stars in his own production. On November 26, 1937, Disney releases "Pluto's Quin-Puplets," in

4. Dave Smith, *Disney A to Z—The Updated Official Encyclopedia* (New York: Hyperion Press, 1998).

which Pluto is left in charge of five puppies as his Pekingese wife, Fifi, goes out for food. The puppies wreak puppy-havoc at home while Pluto gets drunk on moonshine. When Fifi returns, they all get kicked out of the doghouse.

Such are the humble beginnings of a cartoon icon.

While there is no unambiguous link between Pluto the Disney character and Pluto the planet, the connection has always been assumed.[5] We can bet that Walt Disney was not thinking about constipation when he suggested the name for Mickey's dog; before the release of "Mouse Hunt," Pluto the planet had already spent a full year wooing the hearts and minds of the American public. Whether or not Walt Disney was thinking about the cosmos when he named his dog is not important here. What matters is that the seeds were sown for planet Pluto to receive a level of attention from the American public that far exceeds its astrophysical significance in the solar system. The *New York Times* science writer Malcolm W. Brown, in a February 9, 1999, article on Pluto, quoted an unnamed astronomer who made a similar observation:

> *If Pluto had been discovered by a Spaniard or Austrian, I doubt whether American astronomers would object to reclassifying it as a minor planet.*

Over the decades to follow, as the size, influence, and wealth of the Walt Disney conglomerate grew, now a $30 billion company, so, too, did the name Pluto in the collective sentiment of Americans. Indeed, the corporation had achieved a kind of control over our Plutonic emotions, leaving me with no choice but to label the Disney empire what it is:

5. Dave Smith, Chief Archivist, Walt Disney Archives, private communication via Richard Vosburgh.

Plutocracy |plū-tä-krə-sē| (noun) Government by the wealthy.
1) a country or society governed in this way.
2) an elite or ruling class of people whose power derives from their wealth.[6]

As a scientist at New York City's American Museum of Natural History, I sustain an osmotic link with colleagues whose expertise draws from the entire animal kingdom. We've got herpetologists, paleontologists, entomologists, and mammalogists, to name a few. So while I cannot claim fluency on all subjects of natural history, I do claim sensitivity. This leads me to ask how it came to be that Pluto is Mickey's dog, but Mickey is not Pluto's mouse.

Something is awry in the taxonomic class of mammals in the Disney universe.

I would later learn that if you are a Disney character who wears clothes, no matter what your species, you can then own pets, who themselves wear no clothes at all, except perhaps for a collar. Pluto runs around naked except for a collar that says "Pluto." Mickey runs around with yellow shoes, pants, white gloves, and the occasional bow tie; The haberdasheral hierarchy is clear.

One never knows fully how and why some words, names, ideas, or objects penetrate culture, while others fade to insignificance. In straw polls that I persistently take of elementary school children, their favorite planet is Pluto, with Earth and Saturn a distant second. At some level of cognition, the simple sound of a word on the ear or an exotic meaning can make or break a word's popularity and prevalence. Among all planet names, for example, Pluto sounds the most like a punch line to a hilarious joke: ". . . he thought he was on Pluto!" And while the names of all other

6. *New Oxford American Dictionary*, 2nd ed. (New York: Oxford University Press, 2005).

Figure 1.8. *The cultural juxtaposition of Pluto the dog and Pluto the planet makes irresistible content for cartoonists.* Top: *Cartoonist Bill Day, of the* Commercial Appeal, *parodies America's ongoing scientific illiteracy.* Bottom: *Pluto, the most misbehaved of all planets, gets sent to the interstellar doghouse in a comic by Dick Locher, of the* Chicago Tribune.

planets are traceable to mythical gods whose talents or powers one might envy, Pluto is, of course, named for the god of a dark and dank residence for the dead. That's just funny.

In most times and at most places throughout history, the greatest measure of cultural penetration comes not from what sociologists discuss but what artists draw. It may be a while, if ever, before we see a Pluto exhibit at New York's Museum of Modern Art, but that doesn't stop the creative urges of comic strip illustrators from comingling the affairs of Pluto with the affairs of state.

Maybe we shouldn't stand in denial of the provinciality of it all. Disney is an American company. Mickey Mouse is cartoon royalty. Pluto is Mickey's dog. Pluto the planet was discovered by a farm boy from middle America, on a search conducted from the mountains of Arizona, initiated and funded by a descendant of blue-blooded Bostonians.

We have further made a cottage industry of memorizing the sequence of planets from the Sun.

My Very Easy Method Just Simplifies Us Naming Planets.
My Very Excellent Mother Just Served Us Nine Pickles.
My Very Educated Mother Just Stirred Us Nine Pies.
My Very Excellent Man Just Showed Us Nine Planets.
My Very Easy Memory Jingle Seems Useful Naming Planets.
My Very Excellent Monkey Just Sat Under Noah's Porch.
My Very Early Mother Just Saw Nine Unusual Pies.
Mary's Velvet Eyes Makes John Sit Up Nice and Pretty.
Mary's Violet Eyes Makes John Stay Up Nights Pondering.
Many Very Eager Men are Just Sissies Under Normal Pressure.
Man Very Early Made Jars Stand Up Nearly Perpendicular.
My Very Elegant Mother Just Sat Upon Nine Porcupines.

For most of these mnemonics, the word substituted for Pluto represents the principal subject of the sentence, leaving the sentence vulnerable to collapse if the P-word ever disappeared.

From the late 1980s onward, the most popular planet mnemonic has been "My Very Educated Mother Just Served Us Nine Pizzas," associating Pluto with pizza, a favorite food in America,[7] especially among schoolchildren. No other mnemonic has come close to its popularity, in spite of the many clever ones that circulate.

On reflection, I may have strongly influenced the choice of the word *pizza* for the planet mnemonic. In my early years of graduate school (begun at the University of Texas at Austin, but finished at Columbia University, in New York City), I had only ever heard Pluto associated with prunes in the mnemonic, which is surely what your educated mother, who is interested in your gastrointestinal well-being, would serve you, not to mention the distant connection prunes have with Pluto Water as a laxative. I dislike prunes but love pizza. Given that Americans eat 100 acres of pizza a day, I am not alone in that sentiment, and I did not worry about how absurd a serving of nine pizzas would be, compared with being served nine prunes. And so, while I was a teaching assistant in Texas, I remember changing "Prunes" to "Pizza" beginning in 1980 for all the large introductory astronomy classes I taught, which totaled thousands of students by the time I left Texas. I also introduced pizza for the planet mnemonic in my first book, *Merlin's Tour of the Universe*, published in 1988. And I have not once heard prunes associated with Pluto since the early 1990s.

The perennial classroom exercise of memorizing planets in sequence from the Sun allowed the enumeration of the nine planets to take on mythical significance in the minds of students and educators alike. Every printed introduction to the solar system, no matter the grade level of the curriculum, began with a list of the nine planets, in order from the Sun, accompanied by a table or diagram of their relative sizes. This tradition became the pedagogical equivalent of eating comfort food. You somehow knew that all was right with the universe as you learned the planetary sequence, with little

7. Americans eat approximately 3 billion pizzas per year, 100 acres of pizza each day, or about 350 slices per second. Mama deLucas, *All About Pizza*, © 2007; http://www.mamadelucaspizza.com/pizza.

Figure 1.9. *Cartoon postcard by Paul McGehee. Although he drew one for each of the planets, Pluto's cultural popularity surpasses them all.*

Pluto rounding out the list of nine. Even the Planetary Society, an organization founded in 1980 by Carl Sagan and two colleagues, Lou Freidman and Bruce Murray (both from NASA's Jet Propulsion Laboratories in Pasadena), chose as its toll-free phone number 1-8 0 0-9 W O R L D S.

Meanwhile, the *Voyager 1* and *Voyager 2* spacecraft, launched in the 1970s but executing their outer-planet flybys in the 1980s, revealed that the moons of Jupiter, Saturn, Uranus, and Neptune may be as interesting as the planets themselves—maybe more so. It was soon clear that the number of intriguing worlds in the solar system vastly exceeds nine, including seven moons that measure larger than Pluto itself: Earth's Moon; Jupiter's Io, Ganymede, Callisto, and Europa; Saturn's Titan; and Neptune's Triton. The grade school tradition to rote memorize planet names (usually one's first encounter with the solar system) unwittingly concealed a staggeringly rich landscape of objects and phenomena.

RICHMOND TIMES-DISPATCH 8/06 BROOKINS
"PLUTO'S A PLANET... NO, IT'S NOT... PLUTO'S A PLANET... NO, IT'S NOT..."

2

Pluto in History

BEFORE THERE WAS PLUTO THERE WAS PLANET X.

Planet X was the "undiscovered" object in the outer solar system whose gravity was needed to fully account for the motions of the known planets. Heard about it lately? Probably not. That's because it's dead. But widespread belief in the existence of Planet X is what led directly to the systematic search and discovery of what would become Pluto.

The rise of Planet X begins with the German-born English astronomer Sir William Herschel, who more or less accidentally discovered the planet Uranus on March 13, 1781. The episode was an exciting moment in eighteenth-century astronomy. Nobody in recorded history had ever actually discovered a planet. Mercury, Venus, Mars, Jupiter, and Saturn can each be seen relatively easily with the naked eye, and all were known to the ancients. The bias against finding additional planets was so strong that Herschel, even in the face of contrary evidence, assumed he discovered a comet. He even titled his discovery paper "Account of a Comet."[8] Other astronomers were in denial as well. Charles Messier, the eighteenth century's king of comet hunting, noted on April 29, 1781, "I am constantly astonished at this comet, which has none of the distinctive characters of comets."[9]

Archival records of star positions show that several observers had seen Uranus before Herschel did, but each one had mistakenly classified the planet as a star. In an embarrassing example from January 1769, the French astronomer Pierre Charles Lemonnier did *not* discover Uranus six times. When Herschel finally noted that the mysterious object moved, the availability of nearly a century's worth of "prediscovery" data on its position in the sky enabled astronomers to calculate its orbit with good precision. Those calculations showed that the object's orderly, near-circular path, far from the Sun, had nothing in common with the eccentric trajectories of all known comets. At this point, you would have had to be both blind and boneheaded to resist calling the new object a planet.

But all was not orderly in the solar system. Uranus was behaving badly. This new planet's trajectory around the Sun was not following the path Newton's law of gravity would have it take after all known sources of gravity were accounted for. Some astronomers suggested that Newton's laws might be invalid at such large distances from the Sun. Not so crazy: under

8. William Hershel, "Account of a Comet," *Philosophical Transactions of the Royal Society of London* 71 (1781): 492.

9. Quoted in *The Herschel Chronicle*, edited by Constance A. Lubbock (New York: Cambridge University Press, 1933), p. 86.

new or extreme conditions, the behavior of matter can, and occasionally does, deviate from the predictions of the known laws of physics. Only if Newton's theory of gravity had been nascent and untested would one have good reason to question it. By the time Herschel had discovered Uranus, Newton's laws were on a 100-year run of successful predictions. Most famous among them was Edmond Halley's predicted return in 1759 of the comet that would be named in his honor.

The simplest conclusion? Something was lurking undiscovered in the outer solar system—something whose gravity was unaccounted for in the expected orbital path of Uranus.

Beginning in the late eighteenth century, the French mathematician Pierre-Simon de Laplace developed perturbation theory, which he published in his influential multivolume treatise *Mécanique Céleste*. Laplace's new math gave astronomers an indispensable tool to analyze the small gravitational effects of an otherwise undetected celestial object. Mathematicians and astronomers across Europe, armed with these new tools of analysis, continued to investigate what might be perturbing Uranus. In 1845, a young, unknown English mathematician, John Couch Adams, approached Sir George Airy, Britain's astronomer royal, with a request that he search the sky for an eighth planet. But neither looking for planets nor following the leads of young, spunky mathematicians were part of the astronomer royal's job description, so Adams's request was dismissed. The next year, the French astronomer Urbain-Jean-Joseph Leverrier independently derived similar calculations. On September 23, 1846, he communicated his prediction to Johann Gottfried Galle, who was then assistant director of the Berlin Observatory. Searching the sky that same night, Galle found the new planet, soon to be named Neptune, within a single degree of the spot Leverrier had predicted.

But once again, all was not orderly in the solar system. Uranus was still behaving badly, although less so now that the gravity from Neptune had been accounted for. Meanwhile, Neptune's orbit had some peculiarities of its own. Could yet another planet be awaiting discovery?

Figure 2.1. *An 1895 portrait of Percival Lowell looking dapper. Lowell, the founder of Arizona's Lowell Observatory, launched the search for Planet X, which led to the discovery of Pluto.*

In his early years, Percival Lowell indulged a fanatical, even delusional fascination with Mars, claiming that intelligent civilizations were in residence there, digging networks of canals to channel water from the polar ice caps to the cities. He imagined a diminishing water supply, leaving them on the brink of extinction, which fed the *War of the Worlds*, Martian invasion fever of the day. But he devoted most of the rest of his life to the search for the object he called Planet X (X for the algebraic unknown)—the mysterious body in the outer solar system that continued to perturb Neptune. By this reckoning, of course, one might have previously identified Neptune as the Planet X to Uranus

All efforts to predict the location of Planet X based on perturbations to

Neptune came up empty. Any discovery would require a large-area survey of the sky.

When looking for a planet, nobody wants to pore over images of the sky that contain countless millions of dots, hoping to spot the one that moved between one photo and the next. Fortunately, an ingenious mechanical-optical device known as a blink comparator would come to the rescue, streamlining the task. Blink comparators exploit the remarkable ability of the human eye to detect change or motion amid an otherwise unchanging field: Place two photographic images of the same section of the sky, but taken at different times, side by side in precise alignment. Next, flash the two images back and forth in rapid succession. Against the background star field, any speck on the two photographs that brightens, dims, or shifts position from one image to the other becomes immediately apparent.

Percival Lowell died in 1916, but Clyde W. Tombaugh would later be hired by the observatory to carry on this arduous search, which led to the discovery of Planet X in 1930. The young fellow had been looking at a pair of photographic plates he took on January 23 and 29 of the region around Delta Geminorum, the eighth brightest star in the constellation Gemini. Tombaugh became the third and last person ever to discover a planet in our very own solar system.

In any well-designed, well-conducted survey, you don't stop just because you've discovered something. By completing the survey, you might discover something else. So for the next thirteen years Tombaugh scoured more than 30,000 square degrees of sky (out of a total of 41,253 square degrees). He didn't find any objects as bright as or brighter than Pluto. But the time wasn't wasted. The survey discovered six new star clusters, hundreds of asteroids, and a comet and would stand for decades as the most thorough search of the outer solar system.

But was newly discovered Pluto the Planet X of everybody's suspicions? Pluto was first presumed to be of commensurate rank in size and mass with Neptune, itself about 18 times Earth's mass. If Pluto were to perturb

Figure 2.2. Clyde Tombaugh, age 22, poses proudly next to his homemade reflecting telescope. Two years later he would discover Pluto.

Neptune with its gravity, as people suspected it was doing, then Pluto must be at least that size. But Pluto's distance was far beyond the power of available telescopes to see anything other than an unresolved point of light. In fact, Pluto's size and mass could only be guessed at based on Pluto's brightness after you make an assumption about how reflective its surface is.

One clever method to estimate Pluto's size, which gets you a little closer to its mass, is to time your observation for when Pluto moves against a background star, temporarily blocking the star's light. When you combine the distance and orbital speed of Pluto with how long the star has dimmed, you can get a good estimate of Pluto's width on the sky. As more and more stars passed nearer and nearer to Pluto, but without any dimming, astronomers were forced to continually downsize previous guesses for how large Pluto really is.

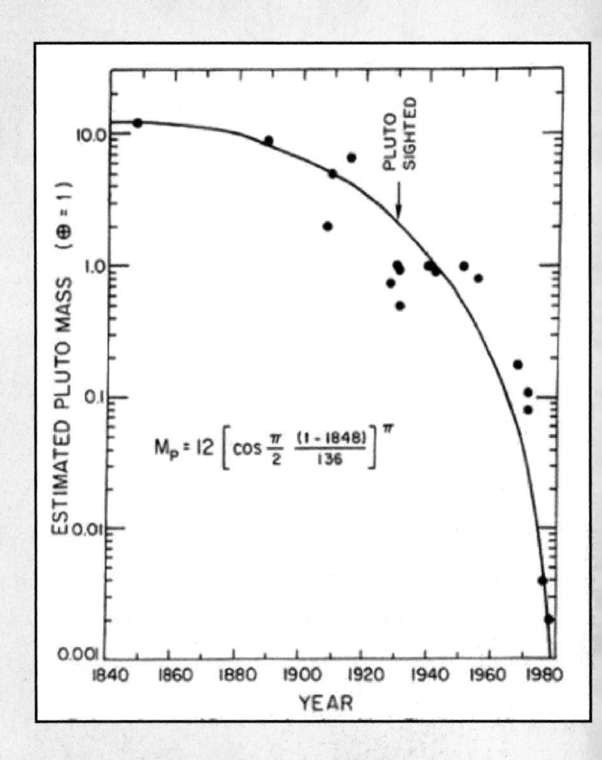

***Figure* 2.3.** *The original plot from Dressler and Russell (1980) in which they show the run of estimates for Pluto's mass, dating back from when it was Planet X. The mathematical equation for* M_P *(the mass of Pluto) is the best-fit model to the data, which carries the distressing news that if the trend continues, Pluto will disappear from the solar system by 1984. (From A. J. Dressler and C. T. Russell, "The Pending Disappearance of Pluto," EOS 61, no. 44 [1980]: 690.)*

In 1978, Pluto was discovered to have a relatively large, close-orbiting moon named Charon, allowing a quality estimate for Pluto's mass. Thanks to a simple application of Isaac Newton's laws of gravity, Pluto dropped precipitously from about Neptune's mass to less than 1 percent the mass of Earth. A 1980 tongue-in-cheek article published in the geology newsletter *EOS* by A. J. Dressler, of Rice University, and C. T. Russell, of UCLA, plotted the mass estimates for Pluto, from its days as Planet X through the 1970s, and predicted that at the rate Pluto's mass was dropping, it would disappear completely from the solar system by 1984 (Figure 2.3).[10]

10. A. J. Dressler and C. T. Russell, "The Pending Disappearance of Pluto," *EOS* 61, no. 44 (1980): 690.

At this level, Pluto's mass was far too small to account for Uranus's and Neptune's orbital oddities. Planet X still had to be lurking, undiscovered, in the outer limits of the solar system.

That was the prevailing belief until May 1993, when E. Myles Standish Jr., of the Jet Propulsion Laboratory in Pasadena, California, published a paper in the *Astronomical Journal* titled "Planet X: No Dynamical Evidence in the Optical Observations." Standish used the updated mass estimates for Jupiter, Saturn, Uranus, and Neptune that had become available from the Voyager flybys; in the case of Neptune, the mass difference amounted to nearly 0.5 percent—quite large by today's standards. Assuming that the masses derived from the Voyager missions were accurate (a wise move), and discounting a single set of suspicious measurements made at the U.S. Naval Observatory between 1895 and 1905 (another wise move), Standish recalculated all the orbital parameters. The result? The misbehaving trends in the paths of Uranus and Neptune disappeared completely, and their orbits could be explained entirely within the gravitational landscape of the presently known solar system. In plain English: Planet X was dead. The inventory of large objects, as decided by the gravity budget of the solar system, was complete.

It seems quite obvious what a planet is, or ought to be. If an object orbits the Sun but is not itself a comet and does not orbit another object the way moons do, then all is well. William Herschel discovered Uranus in 1781. And Johann Galle, of the Berlin Observatory, discovered Neptune in 1846. But few people know that on January 1, 1801, the Italian astronomer Giuseppi Piazzi discovered the planet Ceres happily and silently orbiting the Sun between Mars and Jupiter. The suspiciously large gap between Mars and Jupiter had finally been filled. But astronomers rapidly determined that Ceres was much, much smaller than any other planet. Then on March 28, 1802, the German astronomer Heinrich

Wilhelm Olbers discovered the planet Pallas in the same orbital zone as Ceres. For these two new planets, William Herschel could not identify a visible surface, even through the optics of his powerful telescopes. Apart from their obvious motion across the field of view, the telescopic appearance of Ceres and Pallas was otherwise indistinguishable from that of a distant star. In an expression of sentiment that echoes modern-day debate over what to call Pluto, Herschel wrote, in an 1802 letter to his friend, physician and scientist William Watson:

> *You know already that we have two newly discovered* celestial bodies. *Now by what I shall tell you of them it appears to me much more poor in language to call them planets than if we were to call a* rasor *a* knife, *a* cleaver *a* hatchet, *&c. They certainly move around the sun; so do comets. It is true they move in ellipses; so we know do some comets also. But the difference is this: they are extremely small, beyond all comparison less than planets. . . . Now as we already have Planets, Comets, Satellites, pray help me to another dignified name as soon as possible.*[11]

In a research paper submitted to the Royal Society the following month, Herschel proposed "star-like" as a descriptor, which in Greek becomes the more familiar "aster-oid."

At 600 miles in diameter, Ceres is dwarfed by Mercury, the reigning smallest planet. But that won't concern us just yet. By 1807, three more of these diminutive planets had been discovered: Pallas, Juno, and Vesta. By 1851, 11 more had been logged, and the solar system's planet count reached 18—duly recorded this way in textbooks of the times. The spate of new planets were all small and traveled in orbits similar in size and location to that of Ceres. By 1853, it was clear that a new class of objects

11. Quoted in Michael Lemonick, *The Georgian Star: How William and Caroline Herschel Revolutionized Our Understanding of the Cosmos* (New York: Atlas/Norton, 2008), p. 144.

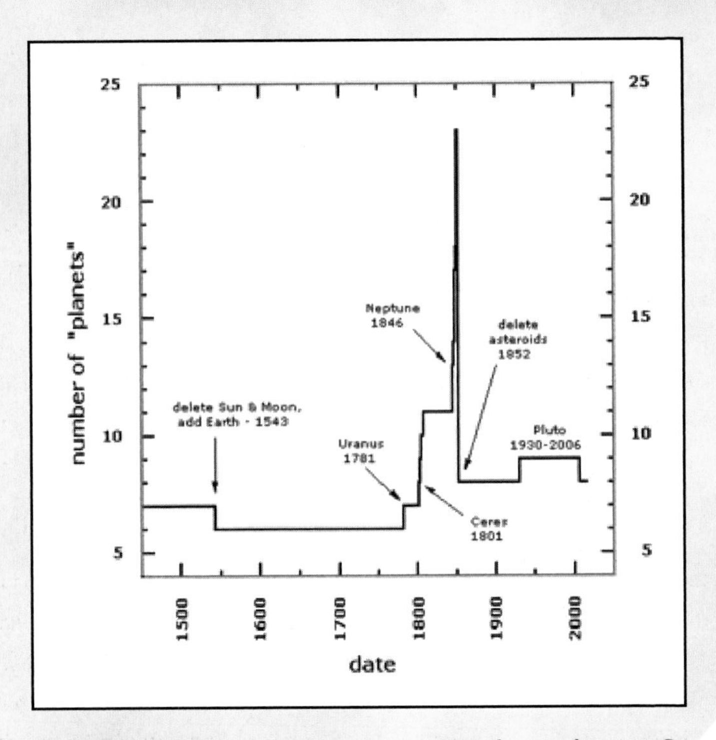

Figure 2.4. The number of planets over time. From the era of ancient Greece until 1543, the number of planets remained unchanged at seven. After Copernicus advanced the Sun-centered system, the number dropped to six, reached a peak of 23 with the discovery of a bunch of asteroids, dropped back to eight when the asteroids became a category of their own, went back up to nine with the discovery of Pluto in 1930, and fell back to eight in August 2006. (Adapted from Steven Soter, "What Is a Planet?" Scientific American, *January 2007.)*

had been identified: *the asteroids*. These bodies occupied a new swath of real estate in the solar system: *the asteroid belt*. Practically overnight, the planet count dropped back to seven: Mercury, Venus, Earth, Mars, Jupiter, Saturn, and Uranus (Figure 2.4).

Ceres was discovered first because it's the brightest and largest of its class. At twice the mass of all other asteroids combined, of which there are hundreds of thousands known, Ceres swiftly went from being the smallest in the class of planet to being the largest in the class of asteroid.

From the time of ancient Greece up to the publication of Nicolas Copernicus's 1543 magnum opus *De Revolutionibus*, the planet count for the known universe was seven. With gods drawn from Roman and Norse mythology, we derive names for the seven days of the week from these objects. The Greeks determined that of all the celestial bodies, only seven move against the background sky. With paths not fully understood, they called this elite group the "wanderers," which in Greek translates to "planets." The planet inventory was clear, clean, and classic: Mercury, Venus, Mars, Jupiter, Saturn, Sun, and Moon.

The Earth-centered view of the universe collapsed after Copernicus, who placed the Sun in the middle, the Moon in motion around Earth, and the Earth-Moon system in motion around the Sun. Thus was born the "solar" system that today we take for granted. But what then becomes of the magnificent seven? The Moon and Sun were pulled from their planet status while Earth joined the list, in commensurate rank with the other revolving objects. This commonsense reassessment of the word *planet* dropped the number to six and did not require a resolution or a formal, agreed-upon definition. It seemed unambiguous enough. Or so we thought.

DON'T UNDERSTAND
OW A SMALL GROUP
F ASTRONOMERS CAN,
N THEIR OWN, DECIDE
THAT PLUTO
ISN'T A
PLANET!
www.foxtrot.com
ALL THE ASTRONOMERS WERE FROM EARTH, TOO! WHO PUT THEM IN CHARGE OF THE WHOLE SOLAR SYSTEM?!
WHERE WERE THE PLUTONIAN ASTRONOMERS?! SHOULDN'T THEY HAVE A VOICE IN SOMETHING LIKE THIS?!
9-14
© 2006 Bill Amend / Distributed by Universal Press Syndicate
TAXONATION WITHOUT REPRESEN-TATION, I SAY!
JUST
AWAY
MY
PLE

3

Pluto in Science

PLUTO CONTAINS ABOUT 70 PERCENT ROCK AND 30 percent ice if you measure things by mass. But rock is denser than ice, leaving the rocky parts to occupy only 45 percent of Pluto's volume. If volume is what matters to you, then you can rightly declare that Pluto is mostly ice. This fact sits within a long list of properties that are not shared with any other planet in the solar system. While each planet is unique in one way or another, Pluto's list just might be as long as all the other planets' combined.

Pluto is by far the least massive planet, with less than 5 percent the mass of Mercury, the solar system's next smallest planet.

Pluto's orbit is so eccentric—so flattened from a perfect circle—that it crosses the orbit of Neptune, its nearest planet (Figure 3.1). In fact, Pluto spends 20 years out of its 248-year orbit closer than Neptune to the Sun.

Pluto's orbit is not only oblong. It tips more than 17 degrees from the plane of the solar system, a full 10 degrees more than Mercury's orbit, the next most tipped in the set (Figure 3.2).

Pluto's largest moon, Charon, named for the Greek ferryboat driver who would carry your unfortunate soul across the River Acheron to the underworld, was discovered in photographs obtained from the Kaj Strand 61-inch telescope at the U.S. Naval Observatory's Flagstaff station in Arizona. It was June 1978 when USNO's James Christy spotted a suspicious bump—an odd elongation to Pluto's grainy image.[12] Richard Binzel (then a graduate student at the University of Texas at Austin) and his collaborators confirmed Charon's existence in February 1985, when the moon eclipsed Pluto, visibly reducing the total light emitted by the system.[13] Pluto had finally reached a place in its orbit where, from Earth's line of site, Charon would pass directly between us and Pluto.

Charon is so large compared with Pluto—equivalently, Pluto is so small compared with Charon—that they orbit a spot that sits not within Pluto itself, but in free space. In other words, unlike the moons of every other planet in the solar system, where their center of motion falls within the physical body of the host planet, the Pluto-Charon system orbits a point in space outside of the physical extent of Pluto itself. Binzel's measurements enabled accurate estimates of Charon's period of revolution

12. J. Christy, "1978 P 1," *IAU Central Bureau of Astronomical Telegrams*, Circular No. 3241, July 7, 1978.

13. R. P. Binzel, D. J. Tholen, E. F. Tedesco, B. J. Buratti, and R. M. Nelson, "The Detection of Eclipses in the Pluto-Charon System," *Science* 228, no. 4704 (1985): 1193–1195.

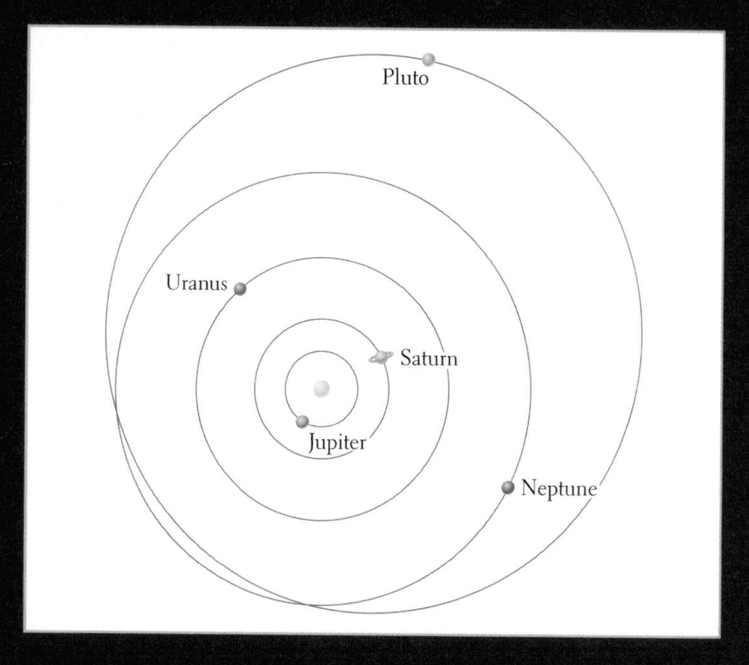

Figure 3.1. *Orbits of the outer planets. When seen from "above," Pluto's orbit is so elongated that it's the only planet to cross the orbit of another planet, in this case Neptune.*

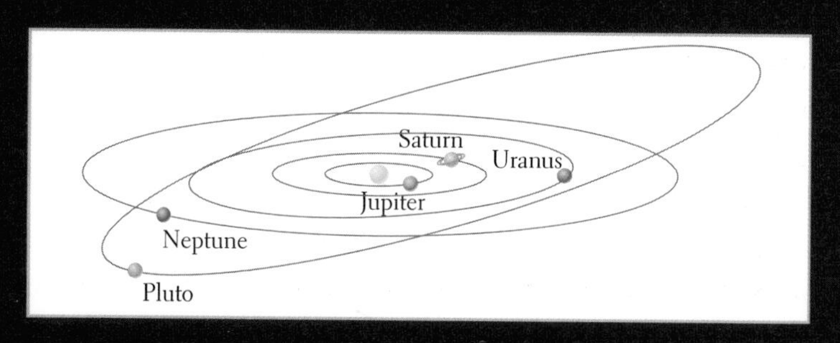

Figure 3.2. *Orbits of the outer planets, in perspective. From this view, the excessive tip of Pluto's orbit from the plane of the solar system is evident.*

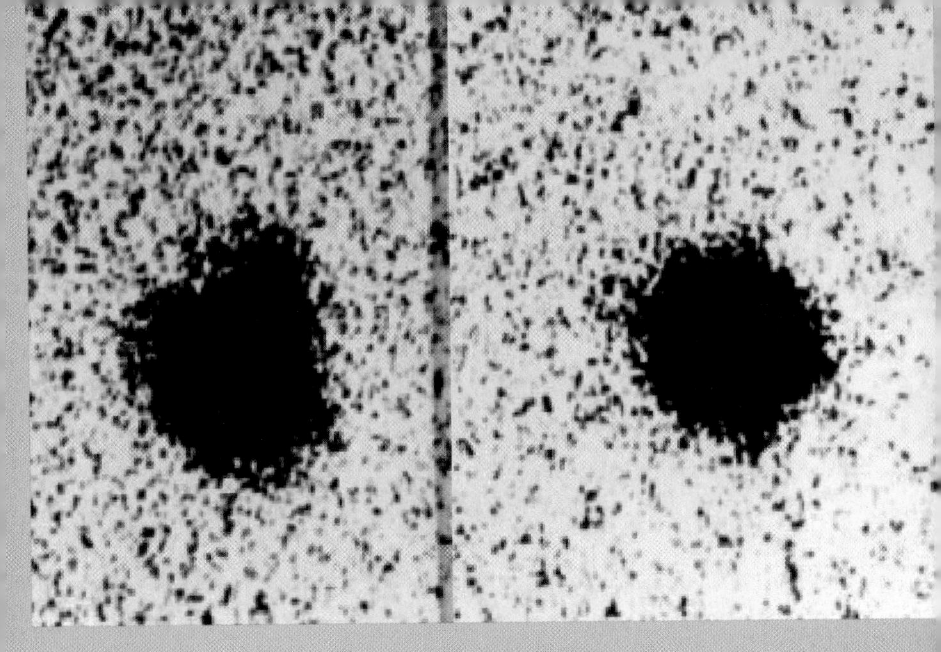

Figure 3.3. Discovery photo (shown in negative) of Pluto's moon Charon. In June 1978, U.S. Naval Observatory astronomer James Christy spotted, in the grainy photo of Pluto on the left, a suspicious elongation to the image. An earlier image of Pluto on the right showed no such distortion, hinting at an orbiting companion to Pluto, which was later confirmed to be a moon.

around Pluto, which perfectly matched the time it took Pluto to turn once on its axis.

Pluto and Charon are in a rare double tidal lock, always showing the same face to each other. Tidal locks themselves are common. Jupiter and Saturn have tidally locked their closest satellites. And Earth has tidally locked the Moon, creating a genuine "near side" and "far side." Although the Moon is trying to tidally lock the Earth, forcing our days to get longer and longer until they equal the lunar month, it will not succeed within the Sun's life expectancy. This leaves Pluto and Charon as the only double-tidally locked system among the planets.

Although Pluto's orbit crosses that of Neptune, they will never collide because Neptune has Pluto in another kind of lock, called an orbital

Figure 3.4. James Christy (seated), the discoverer of Pluto's moon Charon, shown in 1978 with his colleague Robert S. Harrington in front of some computers that were surely considered fast in their day. Harrington would calculate that a series of eclipses between Pluto and Charon might occur in the 1980s.

resonance. For every three trips around the Sun that Neptune takes, Pluto takes exactly two, leaving them in an eternal three-to-two resonance with each other. No other pair of planets behaves this way.

Pluto is further distinguished by the presence of other small icy objects that share its orbital space. If you add up all the debris—leftovers from the formation of the solar system—that can collide with Pluto, it rivals the mass of Pluto itself. All other planets handily dominate their zones. While they (Earth included) continue to be hit by wayward comets and asteroids, the total mass of these impactors remains small compared with the planet itself.[14] Italian astronomer

14. Steven Soter, "What Is a Planet?" *Astronomical Journal* 132 (2006): 2513–2519.

Figure 3.5. Cartoon by Italian astronomer and asteroid hunter Vincenzo Zappalá, poking fun at Pluto and its debris-filled orbit around the Sun.

and asteroid hunter Vincenzo Zappalá could not resist creating a cartoon about it, with Pluto steeped in anthropocentric household trash as the rest of the planets look on in disapproval.

Alas, both Pluto and Charon share an important physical property with the rest of the planets. They're both round. Apart from crystals and broken rocks, not much else in the cosmos naturally comes with sharp angles. The list of round things is practically endless and ranges from simple soap bubbles to the entire observable universe. Spheres tend to take shape from the combined action of simple physical laws. You can show, using freshman-level calculus, that the one and only shape that has the smallest surface area for an enclosed volume is a perfect sphere. In fact, billions of dollars could be saved annually on packaging materials if all shipping boxes and all packages of food in the supermarket were spheres. For example, the contents of a big box of Cheerios would fit easily into a spherical carton that had a 4-inch radius. But practical matters prevail. Spheres don't pack or stack well and nobody wants to chase packaged goods down the aisle after it rolls off the shelves. We already do this for apples and oranges.

For cosmic objects larger than a particular threshold, energy and gravity conspire to turn them into spheres. Gravity is the force that serves to collapse matter in all directions, filling in the low places with material from high places. But gravity does not always win; the chemical bonds of solid objects are strong. The Himalayan range in Tibet continues to grow against the force of Earth's gravity. But before you get excited about Earth's mighty mountains, you should know that the spread in height from the deepest undersea trenches to the tallest peaks in the world is about a dozen miles, but Earth's diameter is nearly 8,000 miles. Contrary to what it looks like to teeny humans crawling on its surface, Earth, as a cosmic object, is remarkably smooth; if you had a gigantic finger and you rubbed it across Earth's surface (oceans and all), Earth would feel as smooth as a cue ball. Expensive globes that portray raised portions of Earth's landmasses to indicate mountain ranges depict a grossly exaggerated reality.

If a solid object has a small enough surface gravity, the chemical bonds in its rocks will easily resist the force of their own weight. When this happens, almost any shape is possible. Two famous celestial nonspheres are Phobos and Deimos, the Idaho-potato-shaped moons of Mars. On 13-mile-long Phobos, the bigger of the two moons, a 150-pound person would weigh only about 4 ounces. All but the largest of the asteroids and comets are too small for their gravity to force them in the shape of a sphere—hence the classic view of craggy chunks of tumbling rock and ice that we carry for these populations.

Our earlier concept of Pluto as a round object, commensurate in other ways with the eight major planets in the solar system, led to the

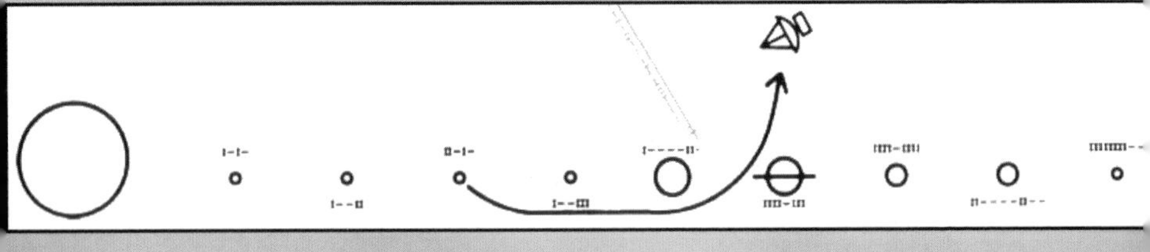

Figure 3.6. *Illustration of our solar system that is sure to confuse aliens. The* Pioneer 10 *and* Pioneer 11 *space probes were launched in the early 1970s and acquired enough energy to escape the gravitation bonds of the Sun. Each craft carried an etched plaque with this iconic view of the planets, intended to alert aliens of the basic structure of our solar system and that the craft itself was launched from the third planet. Given the size of Jupiter and Saturn as illustrated, the scale requires that Pluto not be a small circle as shown, but a pinprick. Furthermore, seven moons that are larger than Pluto are absent from the view, and all four gas giants (Jupiter, Saturn, Uranus, Neptune) have rings, not just Saturn. So an actual alien who used this map in search of our solar system would surely pass us by, certain our map was for some other star system and not the Sun's.*

illustration of the nine-planet solar system, etched on a gold plaque, that accompanied *Pioneer 10* and *Pioneer 11* to the outer solar system. Launched by NASA in the early 1970s, when Pluto's family status was rarely questioned, the *Pioneer* spacecraft were the first objects to ever achieve escape velocity for the solar system. They would never come back. Not ever. And so the schematic view (Figure 3.6) was intended to alert curious aliens from other star systems to the basic form of our solar system and to offer our return address, with a line clearly emanating from Earth, the third planet from the Sun. What the etching did not convey was that given the size of Jupiter and Saturn as illustrated, the size of Pluto would be a pinprick. The plaque further does not show the seven moons of the solar system that are bigger than Pluto.

And unlike what is illustrated, with Saturn's ring clearly drawn, the rest of the gas giant planets (Jupiter, Neptune, Uranus) bear rings as well. So spacefaring aliens would indeed be confused by our illustration and would surely pass us by in search of an actual star system that looks like what we drew.

The curious aliens would also be surprised to learn that on Pluto you would weigh about 10 pounds, while on Jupiter's (undrawn) four largest moons you would weigh upward of 20 pounds. In each of these cases, however, the gravity field sits well above the threshold to overcome tendencies to look like potatoes. Pluto and Charon are in good company, but that company is large and includes nearly all moons, all planets, and all stars.

With an average orbital distance of 40 times that of the Earth from the Sun, Pluto is far. With an average high temperature of –365°F, Pluto is cold. With a diameter smaller than the distance from San Francisco to Topeka (and with Charon checking in at still half that width), Pluto is small. And since no probe has ever visited Pluto, Pluto remains among the least known objects in the solar system. But all that will soon change. After a decade of fits and starts in Congress, the New Horizons flyby mission to Pluto is on its way (Figures 3.7 and 3.8).

The fastest hunk of hardware ever launched, the *New Horizons* spacecraft left Cape Canaveral, Florida, on January 19, 2006, atop the powerful Atlas V rocket. After the second- and third-stage rockets fired, the half-ton, piano-sized spacecraft was endowed with enough speed to pass the Moon's orbit in 9 hours (the Apollo astronauts took 3½ days) and to reach Jupiter for a gravity assist in just over a year. After Jupiter's gravity assist, the craft will be traveling at a breakneck 53,000 miles per hour, nearly 15 miles per second.

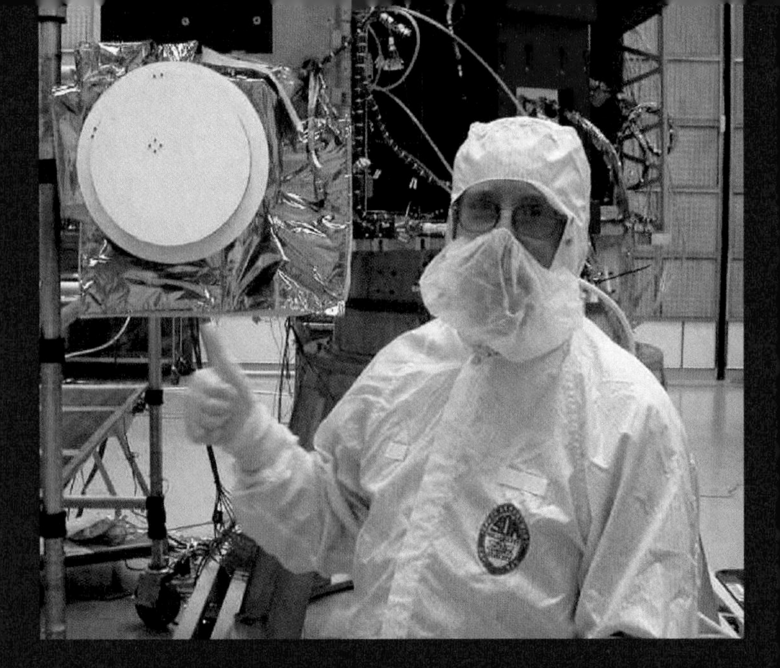

Figure 3.7. *Masked MIT planetary scientist Richard Binzel gives a thumbs-up for the* New Horizons *spacecraft, shown behind him. The craft is in a clean room at the Applied Physics Laboratories of Johns Hopkins University, being prepared for launch. Binzel appears unmasked in Figure 3.10.*

Figure 3.8. *Patch for the* New Horizons *mission, a joint venture of the Southwest Research Institute,* NASA*, and the Johns Hopkins University's Applied Physics Laboratory. Notice that it has nine sides—one of many occasions where the number nine appears in mission materials. A coincidence? Or a subliminal ploy to sway the emotions of the viewer?*

Figure 3.9. Alan Stern (left), principal investigator of the New Horizons mission to Pluto, poses next to the author, both in a desperate attempt to show that astrophysicists can be cool. Photographed at Kennedy Space Center just before the scheduled launch of the New Horizons mission to Pluto, January 2006.

Figure 3.10. James Christy (left), the discoverer of Pluto's moon Charon, stands with Richard Binzel, the first person to measure Charon passing in front of Pluto, which allowed significant conclusions to be drawn about the Pluto-Charon orbiting system.

The lead scientists packed the *New Horizons* spacecraft with seven scientific experiments to answer fundamental questions, such as: What is Pluto's atmosphere made of, and how does it behave? What does the surface of Pluto look like? Are there big geological structures? How do particles ejected from the Sun (the solar wind) interact with Pluto's atmosphere? How empty of dust is the space between Earth and Pluto?

I was kindly invited to the launch by Southwest Research Institute's Alan Stern (Figure 3.9), a Pluto expert, the mission's principal investigator, and a lifelong Plutophile. A magnanimous gesture on Alan's part, knowing my spotty public position regarding Pluto's planethood. I was honored to be asked and I gladly accepted. Also in attendance at Kennedy Space Center that day were the discoverers of Pluto's moon Charon, James Christy and Richard Binzel, as well as Bill Nye the Science Guy® (Figures 3.10 and 3.11). Bill was a student at Cornell when Carl Sagan was a professor there. While known primarily for his expositions of everyday science, the classes he took with Carl imprinted him with a love of the solar system and the rest of the universe that persists to this day.

One of the stated goals for the New Horizons mission was to "complete the reconnaissance of the solar system." This marching order never sat well with me. It conveys a needless tone of finality to the mission. One can just as easily assert that we are "beginning the reconnaissance of a new part of the solar system, previously unvisited," as I consistently conveyed in my public appearances.

In the meantime, the Hubble Space Telescope, famous for its detailed, high-resolution images of gossamer gas clouds across our Milky Way galaxy and of galaxies that reside in the distant universe, was trained on the surrounding environs of Pluto itself. This enabled the Pluto Companion Search Team, led by Hal Weaver and Alan Stern (seen in Figure

Figure 3.11. Like the Academy Awards, but for science, celebrities abound at the January 2006 New Horizons launch to Pluto. One of the nation's leading educators, Bill Nye the Science Guy® (left) poses with the author. In this rare photo, Bill Nye appears without his trademark bow tie. And the author wears an embarrassingly loud necktie that displays eight planets in full view, with Pluto buried in the knot.

3.9), to discover in June 2005 two additional moons in orbit around Pluto (Figure 3.13). A year later, the International Astronomical Union (IAU) officially named them Nix (or Pluto II, the inner of the two moons) and Hydra (or Pluto III, the outer moon).[15]

15. International Astronomical Union, IAU Circular No. 8723, June 21, 2006.

Figure 3.12. Four *onlookers stand agape at the Atlas V rocket, posed to launch the* New Horizons *spacecraft on a fast track to Pluto and beyond. The bulbous nose cone contains the space probe itself. Everything else—the copper cylinder and the white, strap-on boosters—are all rocke fuel.* New Horizons *successfully launched on January 19, 2006. A a peak speed of about 35,000 miles per hour (10 miles per second), it is the fastest-travelin spacecraft ever sent anywhere.*

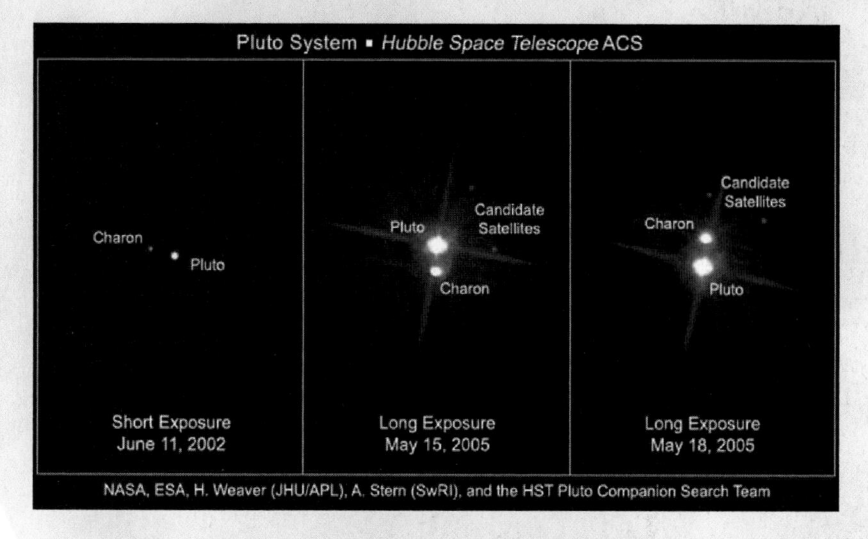

Figure 3.13. *Hubble Space Telescope images of two additional moons of Pluto, visible in each of the long-exposure photographs. Here identified as "candidate satellites," but later confirmed and named Nix and Hydra, they were discovered in orbit around Pluto, allowing Plutophiles the world over to now refer to the "Pluto system" of host planet and three moons, Charon included. Alan Stern (see Figure 3.9), one of the world's leading Pluto researchers, and Hal Weaver were the lead members of the Pluto Companion Search Team responsible for the discovery.*

The amount of brain energy invested in these names knew no bounds. The first letters of the two new moons, N and H, offer respect to the New Horizons mission to Pluto, echoing the fortuitous coincidence that the first two letters of the name Pluto offer tribute to Percival Lowell. Remembering that the Greek beast Hydra sports *nine* heads offers a battery of nods to Pluto's 76-year tenure as the ninth planet. Meanwhile, Hydra's first letter H honors the Hubble Space Telescope, used for the moons' discovery. Nix (the Egyptian variant of Nyx) is named for the Greek goddess of darkness and night. She also happens to be mother of Pluto's other moon Charon, creating a happy orbiting family in the depths of space that many justifiably call the "Pluto system."

MARGULIES
© 2006 THE RECORD NEW JERSEY
Look who they're trying to vote off 'Dancing with the Stars...'
PLUTO

4

Pluto's Fall from Grace

As already noted, our best estimates for the mass of Pluto had been shrinking since the day it was discovered.

Up through the 1970s, the typical astronomy textbook began with a section called "The Solar System" that would profile, chapter by chapter, each planet in sequence from the Sun, ending with Pluto. This approach presupposes that enumerating the nine planets, in order from the Sun, is of fundamental interest and of scientific importance and that it's worthwhile for students to memorize their names in "My-Very-Educated-Mother . . ." order. But by the 1980s, as we discovered more and more comets, asteroids, and moons, and as we continued to characterize their detailed properties, it became clearer and clearer that the Sun's planets are only part of the solar system's story. The outward-bound *Voyager 1* and *Voyager 2* spacecraft, both launched in 1977, played a starring role in that drama. As they separately approached Jupiter in 1979 and the rest of the outer planets over the decade to follow, one of the welcome surprises was that the outer planets' moons turned out to be as fascinating as the planets themselves—maybe more so. Beginning with the Voyager era, objects other than planets began to enjoy their rightful day in the Sun.

As a direct consequence, textbooks began to organize the solar system into scientifically suggestive categories; Pluto, the comets, the asteroids, and other small bodies with interesting features, such as the moons of the outer planets, became parts of chapters whose titles featured words such as "Debris," "Interlopers," and "Vagabonds." This grouping—or regrouping—began in the late 1970s and persisted through the 1980s. Gradually, Pluto and its properties were being taught differently from the rest of the planets in the solar system.

After Pluto's moon Charon was discovered in 1978, defensive Plutophiles were keen to note that only planets have moons, clearly distinguishing Pluto from comets and asteroids. Of course, Mercury and Venus do not have moons, and nobody was rushing to reclassify them. Clearly, then, a planet does not lose status for not having a moon—but surely if an object has a moon, what else could it be but a bona fide planet? Even Merlin, my pen name for two question-and-answer books on the universe, valued this distinction:

Dear Merlin,

What is Pluto, a planet, planetoid, or comet? How will it be determined if Pluto should be demoted to asteroid status?

Roy Krause

Shaw AFB, South Carolina

> Merlin has noticed over the years that many people would like to demote Pluto to an "-oid" status.
>
> But Pluto is twice the size of Ceres, the largest known asteroid, and 50 times the size of the largest comets. When we consider that Pluto has a satellite of its very own it certainly gets Merlin's vote for full rank and privileges of "planet".[16]

Plutophiles grabbed for Pluto's moon as immediate proof of Pluto's planethood—a criterion invented more or less on the spot, but which left them at risk that we might one day discover a moon around an asteroid. What do you do when that happens? This conundrum reveals a deeper truth in science: When your reasons for believing something are justified ad hoc, you are left susceptible to further discoveries undermining the rationale for that belief.

Sure enough, on February 17, 1994, the *Galileo* space probe opportunistically imaged the aseroid Ida while en route to Saturn. While examining the data, mission member Ann Harch discovered that Ida has a small (1.4-kilometer) orbiting moon that came to be called Dactyl. Idaho-potato-shaped Ida is only 30 miles long and about 12 miles across (Figure 4.1). The thing is unimpeachably asteroidal. And in the time since Dactyl's discovery, detailed observations of many more asteroids suggest that asteroid moons are common. Furthermore, some asteroids may not be solid at all. Many are composed of loosely assembled rubble, some the size of Dactyl itself, which undermines the concept of moon altogether.

16. Neil deGrasse Tyson, *Merlin's Tour of the Universe: A Skywatcher's Guide to Everything from Mars and Quasars to Comets, Planets, Blue Moons, and Werewolves* (New York: Main Street Books, 1997), p. 62.

Figure 4.1. Galileo *spacecraft image of asteroid Ida taken 14 minutes before closest approach in 1993. Ida's tiny moon Dactyl orbits a short distance away to its right. At a mere 30 miles long and with an irregular potato shape, Ida is clearly an asteroid with a moon, undermining the moon criterion for planet status among Plutophiles after Pluto's moon Charon was discovered in 1978. (NASA Jet Propulsion Laboratory Planetary Photojournal; http://photojournal .jpl.nasa.gov/jpeg/PIA00069.jpg.)*

Two kinds of scientists populate the world: those who see what is similar among objects and explore how they differ from one another, and those who see what is different among objects and explore how they're all similar. To arrive at a deep understanding of the natural world often requires a sustained but resolvable tension between the two camps. Even after the shift in the 1980s, Pluto was still a planet, by anybody's reckoning. But behind closed doors, planetary geologists recognized that Pluto possessed many properties that resemble those of comets and asteroids.

This new approach to teaching the solar system didn't just affect Pluto. The rest of the planets were grouped as well. Mercury, Venus, Earth, and Mars became the terrestrial planets, treated as a conceptually coherent subject: they are all small, rocky, and dense. Meanwhile, Jupiter, Saturn, Uranus, and Neptune became the Jovian planets, all of which are large, ringed, gaseous, low density, and fast rotating. Meanwhile, the rest of the Astro 101 textbook began to fill up with discoveries regarding the Big Bang, galaxy formation, galaxy collisions, black holes, the births and deaths of stars, and the search for life.

The astrophysics community, primarily the planetary scientists, simply shifted how they thought about the contents of the solar system. Of course, Saturn is still very different from Jupiter, and Earth is very different from Venus. But Earth and Venus have much more in common with each other than either Earth or Venus has with Jupiter or Saturn. And Jupiter and Saturn have much more in common with each other than either they or the terrestrial planets have with Pluto. With Pluto's properties (size, orbit, composition) sitting alone among the planets, can you justify a class of one? No. Classification schemes require at least two similar objects to define a class. Until that happens, you must find something else to do with your unusual object.

Yes, in a sense, Pluto had no class. But that would shortly change.

In 1992, University of Hawaii astrophysicist David Jewitt and his graduate student Jane Luu used their 2.2-meter optical telescope at Mauna Kea to discover an icy object, cryptically labeled 1992 QB1,[17] orbiting the

17. In the official labeling scheme for newly discovered objects in the solar system, before they are formally named, the first four numerals (1992) are the year of discovery. The first letter (Q) references the semimonth of discovery. Omitting the I and Z, 24 letters remain to span each semimonth of the calendar year. The next letter (B) corresponds to the numerical sequence in which the object was discovered during that half month. Omitting I, this scheme allows up to 25 objects to be discovered in any half month. But if more than that are discovered—a common occurrence these days—then a numeral is appended (1) and the letters repeat. So, in fact, 1992 QB1 was the 27th object discovered in the first half of September in the year 1992.

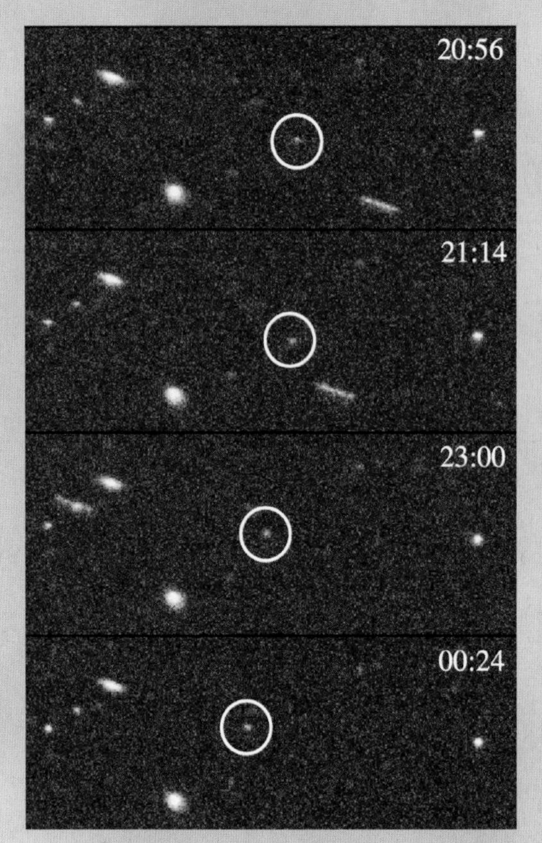

Figure 4.2. Discovery images from 1992 of the first icy body discovered in the outer solar system since Pluto in 1930. Taken by University of Hawaii astrophysicist David Jewitt and his graduate student Jane Luu, using the 2.2-meter telescope on Mauna Kea, this object, labeled 1992 QB1 and identified by the shifting arrow, was the first of many to be discovered in a new region of the solar system called the Kuiper belt.

Sun out beyond Neptune, just where (forty years) earlier the University of Chicago planetary astronomer Gerard Kuiper had hypothesized such objects would live.

One of the biggest problems you face when observing objects in the solar system is that they don't radiate their own light. The most distant ones are so far away, the feeble sunlight that reaches them must then reflect from their surface and make it all the way back to the inner solar system before reaching our telescopes here on Earth. A reflective surface helps. Out there in the cold depths of space, clean ice can satisfy this need. But nobody knew for sure what 1992 QB1 was made of. All they knew was that it orbited the Sun beyond Neptune and that it was small, maybe a fifth the size of Pluto. 1992 QB1 was not a threat to Pluto's regional prominence, but it nonetheless made you go "hmmm."

Jewitt and Luu looked for more. And they found more. One after another after another, all with orbits a bit tipped out of the plane of the solar system, just like Pluto's, some with orbits so elongated that they crossed the orbit of Neptune, just like Pluto. This growing family of objects populated a new swath of real estate that orbits the Sun—a belt, analogous to the band of objects between Mars and Jupiter that came to be known as the asteroid belt.

Gerard Kuiper had proposed that beyond the outermost planet in the solar system (perhaps in any star system) lies a reservoir of slowly orbiting debris—leftovers from the formation epoch that never got "vacuumed" up by a planet's gravity or, more importantly, never coalesced to form a planet in the first place. By comparison, planetary orbits from Mercury to Neptune are relatively free of debris. Even though Earth plows through hundreds of tons of meteors a day—the source of our nightly display of shooting stars—this pales compared with what floats in the outer solar system. So imagine how much sweeping up a massive planet in the outer reaches could do. But once you pass Neptune, you've run out of big planets. And there's so much space among the debris that it all stays there, in orbit, and in vast quantities.

At those distances, 5 billion miles from the Sun, temperatures dip below –400°F and stay there. Plenty of cosmically common ingredients that would evaporate when brought close to the Sun, such as water, carbon dioxide, ammonia, and methane, stay forever frozen at those temperatures, becoming a basic constituent of solid matter. Within a few years of Jewitt and Luu's discovery of 1992 QB1, enough additional objects had been found to confirm that the solar system indeed contains a "Kuiper belt" of icy bodies. Looking at the distribution of sizes found and the rate at which these objects were being discovered, astrophysicists knew that it was just a matter of time before hundreds or even thousands of Kuiper belt objects would be discovered and cataloged. Makes you wonder:

***Figure 4.3.** Clyde Tombaugh, discoverer of Pluto, seen here at age 90, the year before his death. Rarely seen without his cane, it served not only as a walking aid but as a means of punctuating his comments about why Pluto should stay a planet forever.*

What happens the day we find something bigger than Pluto? Do we call it a planet, because Pluto is a planet, or do we use the opportunity to come up with modified nomenclature for this new class of objects, including Pluto?

Clyde Tombaugh was still alive in the early 1990s. He saw the Kuiper belt omens, but fought them tooth and nail with cane in hand, using his cane not only as a walking aid but also as punctuation for the aggressive arguments he would make. Tombaugh had the most to lose if Pluto were classified as anything other than a full, red-blooded planet. In a December 1994 letter to the editor of *Sky & Telescope* magazine (the monthly bible for amateur astronomers), Tombaugh declared:[18]

18. Clyde Tombaugh, "The Last Word," Letters to the Editor, *Sky & Telescope*, December 1994, p 8.

> *I'm fascinated by the relatively small "ice balls" in the very outer part of the solar system. I have often wondered what bodies lay out there fainter than 17th magnitude, the limit of the [photographic] plates I took at Lowell Observatory. May I suggest we call this new class of objects "Kuiperoids"?*

Not knowing that objects larger than Pluto awaited discovery in the Kuiper belt, Tombaugh was unwittingly suggesting that Pluto become a Kuiperoid as well. In any case, astronomers are not likely to adopt a word that sounds like a contagious skin disease.

Clearly frustrated by all the talk of reworking a time-honored classification scheme, Tombaugh proceeds to attack other astronomical traditions that carry historical concepts into the present, including our entrenched and arcane classification system for the spectra of stars:

> *While we are considering reclassifying astronomy, how about revamping the Hertzsprung-Russell diagram so the spectral types [of stars] are alphabetically ordered? No, that would wreck extensive catalogs of stellar spectra. Or let's throw out the awkward constellation system! Alas, that would discard our beautiful mythology.*

Tombaugh now raises "cane" as he goes in for the kill:

> *Pluto started out as the ninth planet, a supported fulfillment of Percival Lowell's prediction of Planet X. Let's simply retain Pluto as the ninth major planet. After all, there is no Planet X. For 14 years, I combed two-thirds of the entire sky down to 17th magnitude, and no more planets showed up. I did the job thoroughly and correctly. Pluto was your last chance for a major planet.*

CLYDE W. TOMBAUGH
Mesilla Park, New Mexico

Who are any of us to argue with the octogenarian discoverer of Pluto?

Just like Paul Revere, John Henry, Paul Bunyan, Davy Crockett, and other folk heroes, Clyde Tombaugh is memorialized in song. Written in 1996 by New York–based singer-songwriter Christine Lavin, "Planet X" (see Appendix B for complete lyrics) is a hilarious, historical account of Pluto, from before it was discovered through Tombaugh's modern efforts to protect its planethood:

It takes 247 earth years
for Pluto to circle our sun.
It's tiny and it's cold
but of all heavenly bodies
it's Clyde Tombaugh's favorite one.
He's 90 now and works every day
in Las Cruces, New Mexico
determined to maintain the planetary status
of his beloved Pluto.

Lavin's 119 lines of text, sung in a kind of folk-rap style, include references to Disney:

That same year, 1930, Walt Disney
debuted his own Pluto as well
but a cartoon dog with the very same name as the CEO of Hell
was not your normal Disney style

To disgruntled horoscope readers:

and Scorpios look up in dismay
because Pluto rules their sign.
Is now reading their daily Horoscope
just a futile waste of time?

And to an empathetic (de-sainted) St. Christopher:

St. Christopher is looking down on all this
and he says, "Pluto, I can relate.
When I was demoted from sainthood
I gotta tell you little buddy,
it didn't feel real great"

Lavin was inspired to write the song after reading an article on Pluto by Sal Ruibal in the March 4, 1996, edition of *USA Today*.

Clyde Tombaugh died January 17, 1997, just a couple of weeks short of his February 4 birthday, when he would have turned 91. He was a leading force for Pluto, but he was not alone in his support of planethood. Behind him were many whose professional research interests focused on Pluto and who wanted to see a mission sent there. By the 1990s, space probes had flown by, or visited, every planet in the solar system but Pluto. Some groups vying for space missions invented catchy slogans to help sell a flyby of Pluto to Congress, like "the first mission to the last planet." Phrases such as that imply (1) that the concept of "planet" is strong and real, (2) that Pluto is a planet, and (3) that once you've been to Pluto, your reconnaissance of the planets is complete. Such paradigms require, of course, that Pluto be a planet. In addition, there was legitimate, if unspoken, concern that if Pluto were demoted to "ice ball," or anything less than planet, then funding for a major Pluto mission could be jeopardized. Why? If Pluto was just a ball of ice, astrophysicists could simply study a passing comet and save the American public the money required to travel 4 billion miles to the outer solar system.

Gerard Kuiper himself voted for demotion of Pluto before anybody else, but for reasons we would deem trivial today. A news story from the science section of *Time* magazine from February 20, 1956, was prophetically titled

Figure 4.4. *Cartoonist Tom Briscoe understood Pluto's needs in times of despair.*

"Demoted Planet."[19] The editors begin bluntly—"Astronomers have always felt uncertain about Pluto . . ."—and go on to list the well-known (oddball) features that distinguish Pluto from the rest of the nine, ending the opening paragraph with, "These deviations suggest that Pluto may not be a real planet." Then, in the next paragraph, *Time* reports Kuiper's additional, but retrospectively lame, argument for demotion:

> *Last week Astronomer Gerard Peter Kuiper (rhymes with piper) of the University of Chicago made another move toward demoting Pluto.*

19. "Demoted Planet," *Time*, February 20, 1956; http://www.time.com/time/archive/preview/0,10987,808181,00.html?internalid=related3.

Recent observations have proved that its period of rotation on its own axis is more than six days. For a planet, says Scientist Kuiper, this is too slow.

Unknown to Kuiper (and to anybody else) at the time, Venus, Earth's "sister" planet, takes 243 days to spin once on its axis, which is 18 days longer than it takes Venus to orbit the Sun itself. In other words, the Venus day is longer than the Venus year, yet nobody is rushing to demote Venus. Mercury has a long day too, lasting two-thirds of its year. Such are the risks of classifying an object on the premise that you have isolated its fundamental features for all time.

After a year of advising the trustees of the American Museum of Natural History on what they might do to reverse the precipitous drop in attendance at its famed Hayden Planetarium, I was appointed acting director. A year after that, in May 1996, museum president Ellen Futter and provost Michael Novacek (a dinosaur paleontologist) formally appointed me as the first occupant of a newly endowed chair, becoming the Hayden Planetarium's ninth director. From day one, my immediate and biggest task was to serve as project scientist for the creation of the museum's new $230 million Rose Center for Earth and Space, named for New York real estate magnate and museum trustee Frederick P. Rose and his wife, Sandra P. Rose, the source of its lead gift and the source of the named academic chair that I occupy. This new facility would contain a freshly conceived and outfitted Hayden Planetarium as part of a huge museum wing dedicated to the universe.

Four principal entities collaborated to shape the look, feel, and content of the Rose Center: (1) the architectural firm Polshek and Partners; (2) the exhibition design firm Ralph Appelbaum and Associates, known to many for their work on the Holocaust Memorial Museum in Washington,

D.C.; (3) the science advisory committee, of which I served as chair, consisting of staff scientists hired for just this purpose:

- James Sweitzer, a University of Chicago astrophysicist turned education professional
- Frank Summers, a Princeton cosmologist
- Steven Soter, a planetary scientist who apprenticed at Cornell University
- Charles Liu, a Columbia University expert on galaxy formation and evolution

as well as selected colleagues drawn from outside the museum with expertise in subfields of astrophysics not represented in the profiles of the local staff; and (4) scientific visualization professionals, led by the astrophysically literate artist Dennis Davidson.

In the old days, a planetarium visit would target the sky show. The exhibits that lined the feeder corridors were what occupied your idle time while waiting for the show to start. By the late twentieth century, however, astrophysicists had compiled much more than a planetarium show's worth of information about the universe. So our task was not to face-lift the existing facility, but to invent something entirely new. Besides designing and acquiring state-of-the-art technology to deliver what we now call space shows, we were constructing a unique and arresting architectural facility that would offer ample three-dimensional exhibit spaces suitable for telling cosmic stories on a grand scale.

The basic architectural design of the Rose Center became a matter of public record in January 1995. It would be a huge, 87-foot-diameter sphere, containing the planetarium space theater within its upper half and another theater in the lower half, featuring a walk through the re-creation of the Big Bang. The entire sphere would be supported from its sides, appearing to float above a sprawling Hall of the Universe below it, all strikingly lit within a cubic glass building and visible from

Figure 4.5. The Frederick Phineas and Sandra Priest Rose Center for Earth and Space, seen at night, containing the Hayden Sphere. This $230 million facility opened to the public on Saturday, February 19, 2000, with solar system exhibits that grouped Pluto with the swarm of icy bodies in the outer solar system known as the Kuiper belt, instead of with the other eight planets of the solar system. This decision made a page 1 story in the New York Times *and angered schoolchildren across the country.*

the street. We spent the next two years establishing our philosophical approaches to the interplay of design and content before we began three years of total reconstruction in January 1997, during which we turned our attention to the exhibit text and other detailed features of the content. Given how common spheres are in the universe, we knew from the start that the Hayden sphere should serve not only as an enclosure but as an element of exhibitry.

To plan content, we first needed to assess the shelf life of various astrophysical subjects. For example, ever since Copernicus, we've been convinced that Earth goes around the Sun and not the other way around.

That would be content of long shelf life that we can boldly cut into metal displays.

In the moderate shelf life category, there's the question of water on Mars. Consensus says that the flowing liquid water that used to be there is currently locked in permafrost, but that notion could get modified by the discoveries of curious rovers on any next NASA mission to the Red Planet. So we display this text and related images with replaceable rear-lit transparencies. Science of possibly brief shelf life would include late-breaking discoveries, any intriguing hypothesis, anything waiting to be verified or trashed by another group of researchers with a different discovery or a more comprehensive theory. For that we simply show videos of research scientists giving their latest ideas. No transparencies. No cut metal. Just swappable video content. Where a given topic landed in these three tiers of information determined the nature of the exhibit treatment it received—which is code for how much money we spent to create the exhibit.

We hired Steven Soter in November 1997, luring him away from the Smithsonian Air and Space Museum. Soter's résumé includes collaborating with Carl Sagan and Ann Druyan on the writing of the landmark PBS series *Cosmos*. Just a few months after his arrival, Steve handed me a February 1998 *Atlantic Monthly* article on Pluto titled, "When Is a Planet Not a Planet?" written by journalist David H. Friedman. At the top Steve politely penned: "Perhaps we should look into this!" He (correctly) figured that the issues raised in the article might influence the content of our planet exhibits, which were still under design.

I decided to write an essay of my own on the subject, which became "Pluto's Honor," for the February 1999 issue of *Natural History* magazine,[20] timed to coincide with the month that Pluto, in its badly elongated trajectory, recrossed the orbit of Neptune after 20 years, once again becoming the farthest planet in the solar system. My intent was not only

20. Neil deGrasse Tyson, "Pluto's Honor," in *Natural History* 108, no. 1 (February 1999): 82.

to celebrate Pluto's regaining its far-out status, but also to review the saga and character of Pluto; raise the historical analogy with the asteroid Ceres, which had been labeled a planet when discovered in 1801; and generally address what was simmering in the minds of planetary scientists. At the end of the article, having laid out the various parameters and arguments, I offered a last gasp of sentiment for the little fellow:

> *As citizen Tyson, I feel compelled to defend Pluto's honor. It lives deeply in our twentieth-century culture and consciousness and somehow rounds out the diversity of our family of planets like the troubled sibling of a large family. Nearly every school child thinks of Pluto as an old friend. And there was always something poetic about being number nine.*

But I could not hold back my fundamental conclusions:

> *As professor Tyson, however, I must vote—with a heavy heart—for demotion. Pluto was always an enigma to teach. But I'd bet Pluto is happy now. It went from being the runt of the planets to the undisputed King of the Kuiper belt. Pluto is now the "big man" on a celestial campus.*

At the same time, I had no intention of unilaterally imposing my personal perspective on the Rose Center's presentation of Pluto. That would be professionally irresponsible and would represent an abuse of my authority as project scientist. To do anything unorthodox with our treatment of Pluto would require a consensus of the internal and external science committee. For one thing, planets are not my research area of expertise. I specialize in star formation and galaxy evolution. For another, it's simply not my role to invent a new classification scheme.

The article triggered a flow of letters to me and to the magazine, the most memorable of which came under the letterhead "Pluto Planetary

Protective Society" and written by its founder and president, Professor Julian Kane, Hofstra University, Long Island, New York. With a clever cue on my "heavy heart" line from the essay, Kane ends his letter with:

> *Professor Tyson maintains that with a heavy heart he must vote to demote Pluto. Professor Kane, however, votes with an atrially-fibrillated heart to sustain Planet Pluto while additional factual details are being uncovered.*

Meanwhile, we at the museum were not the only ones grappling with classification schemes of objects in the outer solar system. The International Astronomical Union (IAU) was, too. Founded in 1919 and with a current membership of 10,000, the IAU is the professional society for all the world's astrophysicists, operating under the mission "to promote and safeguard the science of astronomy in all its aspects through international cooperation." Among its many duties, the IAU establishes committees and other consensus-building activities to formalize our occasionally confusing nomenclature and lexicon. Their authority derives not from law or dogma but from emergent scientific consensus. Not blind to the Pluto–Kuiper belt ruminations of its membership, they decided to look into what was happening out there. This simple and normal step for them was widely regarded by the media (and by many in the community of astrophysicists) as the IAU raring to demote Pluto. Many in the planetary science community were outraged, fearful that the Plutophiles among them might not fully vet their views.

Coincidentally, the same month that my article on Pluto appeared in *Natural History* magazine, IAU general secretary Johannes Andersen issued a candid but clumsy press release denying any rumors that the IAU had endorsed a plan to demote Pluto,[21] which is produced here in its entirety:

21. International Astronomical Union, Press Release 01/99 from the General Secretary, February 3, 1999.

THE STATUS OF PLUTO: A CLARIFICATION

Recent news reports have given much attention to what was believed to be an initiative by the International Astronomical Union (IAU) to change the status of Pluto as the ninth planet in the solar system. Unfortunately, some of these reports have been based on incomplete or misleading information regarding the subject of the discussion and the decision making procedures of the Union.

The IAU regrets that inaccurate reports appear to have caused widespread public concern, and issues the following corrections and clarifications:

1: No proposal to change the status of Pluto as the ninth planet in the solar system has been made by any Division, Commission or Working Group of the IAU responsible for solar system science. Accordingly, no such initiative has been considered by the Officers or Executive Committee, who set the policy of the IAU itself.

2: Lately, a substantial number of smaller objects have been discovered in the outer solar system, beyond Neptune, with orbits and possibly other properties similar to those of Pluto. It has been proposed to assign Pluto a number in a technical catalogue or list of such Trans-Neptunian Objects (TNOs) so that observations and computations concerning these objects can be conveniently collated. This process was explicitly designed to not change Pluto's status as a planet.

A Working Group under the IAU Division of Planetary Systems Sciences is conducting a technical debate on a possible numbering system for TNOs.

Ways to classify planets by physical characteristics are also under consideration. These discussions are continuing and will take some time.

The Small Bodies Names Committee of the Division has, however, decided against assigning any Minor Planet number to Pluto.

3: From time to time, the IAU takes decisions and makes rec-

ommendations on issues concerning astronomical matters affecting other sciences or the public. Such decisions and recommendations are not enforceable by national or international law, but are accepted because they are rational and effective when applied in practice. It is therefore the policy of the IAU that its recommendations should rest on well-established scientific facts and be backed by a broad consensus in the community concerned. A decision on the status of Pluto that did not conform to this policy would have been ineffective and therefore meaningless. Suggestions that this was about to happen are based on [an] incomplete understanding of the above.

The mission of the IAU is to promote scientific progress in astronomy. An important part of this mission is to provide a forum for debate of scientific issues with an international dimension. This should not be interpreted to imply that the outcome of such discussions may become official IAU policy without due verification that the above criteria are met: The policy and decisions of the IAU are formulated by its responsible bodies after full deliberation in the international scientific community.

Johannes Andersen
General Secretary, IAU

If Anderson's point was simply to dispel misinformation, the task required many fewer words than what appeared. The release reads as if written by a lawyer instead of scientist. And from its tone and blatantly defensive posture, methinks the gentleman did protest too much, implying self-awareness of a rising storm.

Given how much money we had spent (and were about to spend) on exhibit design and content, it was incumbent on us not to make hasty decisions about anything astrophysical. We needed to assess, to the best of our abilities, the trends in cosmic discovery, so that long after opening day our exhibits would remain as fresh as possible. So I organized and hosted a panel debate on Pluto's status, inviting the world's leading thinkers on the subject to duke it out, on stage, for our benefit and for the benefit of the interested public.

Eight hundred people descended on the main auditorium (which doubles as an IMAX theater) of the American Museum of Natural History on Monday evening, May 24, 1999, to hear "Pluto's Last Stand: A Panel of Experts Discuss and Debate the Classification of the Solar System's Smallest Planet." You couldn't get more expert experts than the five scientists who joined me onstage:

- Michael A'Hearn, a comet and asteroid specialist at the University of Maryland in College Park, president of IAU's Planetary Systems Sciences Division, and chair of IAU's Committee for Small Body Nomenclature
- David H. Levy, patron saint of amateur astronomers worldwide, discoverer or codiscoverer of dozens of comets and asteroids, and biographer of Clyde Tombaugh
- Jane Luu, professor at Leiden University in the Netherlands and codiscoverer not only of the first actual Kuiper belt object but of many others that followed
- Brian Marsden, a comet and asteroid specialist at the Harvard-Smithsonian Center for Astrophysics and director of IAU's Minor Planet Center and the Central Bureau for Astronomical Telegrams, a clearinghouse for discoveries and transitory astronomical phenomena
- Alan Stern (see Figure 3.9), of the Southwest Research Institute in Boulder (later to become associate administrator for science at NASA), specialist in everything small in our solar system, author of the wonderfully titled *Pluto and Charon: Ice Worlds on the Ragged Edge of the Solar System*,

> and the person who would become principal investigator of NASA's New Horizons mission to Pluto and the Kuiper belt.

If any set of people was going to give us insight into the Pluto problem, it would be them. The right people at the right time and at the right place.

After I briefly reviewed the scientific and pedagogical challenges we all faced, each panelist opened with introductory remarks before the roundtable discussion began. During the 90-minute panel, A'Hearn came off as the strictly-business Pluto rationalist, willing to let you call Pluto a planet if you were investigating the inner working of sizable round bodies or to call it a Kuiper belt object (KBO) if you were investigating its origins. Levy was the unabashed sentimentalist—about Pluto, about astronomy, about science in general. Luu, supremely articulate and confident and by far the youngest person on the stage, was the dispassionate Pluto denigrator and rejectionist, terminally bored by what she considered an old and irrelevant question. Marsden, with his British accent, was the witty, affable multicategorizer. Stern took the Pluto-is-a-planet posture, simultaneously committing to the verdicts of both the laws of physics and the feelings of fifth graders. For that evening, I was the dispassionate facilitator, inwardly open to arguments on all sides even though I'd already declared a preference in print.

Jane Luu kicked off the proceedings—and wasted no time getting her point across:

> *When Jewitt and I discovered the Kuiper belt and showed that Pluto was just another member of [it], we were quite pleased. . . . But I found out that our discovery raises questions that have proved to be painful for many people: Is Pluto truly a planet? Some scientists assert that Pluto's tiny size and its membership in a swarm of*

> *similar objects mean that it should be classified a minor planet, like asteroids and comets. Others are outraged by the idea, insisting that regardless of how its identity has changed, demoting Pluto would dishonor astronomical history and confuse the public.*
>
> *I personally don't care one way or the other. Pluto just goes on the way it is, regardless of what you call it.*

If that was not enough, Luu next drew the battle line between science and sentiment:

> *If Pluto continues to be referred to as the ninth planet, it would only be due to tradition and sentimental reasons. People are fond of planets, because the idea of a planet conjures up notions of home, life, happy things, and astronomers are always looking to find more planets, not to lose them. So in the end, the question goes back to this: Should science be a democratic process, or should logic have something to do with it?*

She also reminded the audience, as I do earlier in this book, that when Ceres was first discovered, it (like Pluto) was thought to be a missing planet, but when other asteroid discoveries quickly followed, astronomers realized that Ceres was just the largest member of a new class of objects in a new place in the solar system—the asteroid belt—and so "its planethood was promptly revoked." Therefore, she contended, if other members of the Kuiper belt had been discovered right after Pluto was, the planethood of Pluto would have been revoked with equal speed.

Luu ended by raising a blindingly obvious question:

> *We are continuing to try to find more Kuiper belt objects, and the search is going pretty well. What if we find other objects fairly close in size to Pluto—maybe even bigger, or maybe just a bit smaller—will these objects also be called planets, or what?*

Her delivery was uncompromising, her tone icy. The audience loved her.

Next up was Alan Stern, on the opposite side of the fence and only slightly more willing than Luu to compromise, but a bit friendlier in tone. Like Luu, he declared that Pluto's planethood should not be a matter of democracy, and like Luu (as well as the rest of the panelists), he chose a clever analogy. Referring to the problem of misconceived terminology, Luu had pointed out that Native Americans had been initially referred to as Indians "simply because Columbus screwed up" and thought he'd arrived in India, but that now we know Native Americans are not the native inhabitants of the Indian subcontinent. Using his own analogy, Stern referred to the alleged problem of Pluto's small size by pointing out that nobody thinks a Chihuahua isn't a dog just because it's small—that there's "something innate" about a Chihuahua, "something doggy" that automatically puts it in the class of dog for any observer. By analogy, Pluto's roundness puts it in the class of planet.

In an effort to frame the ultimate definition of a planet—an effort that the IAU had been studiously avoiding—Stern presented the audience with a "rational sieve," a physical test that anyone in the audience would be able to apply to any celestial object, whether already known or yet to be discovered. Its ease of application, he said, derived from its having an upper size limit and a lower size limit. At the upper end, any object big enough to fuse hydrogen in its interior is acting like a star and must be called a star. At the lower end, any object that doesn't have enough mass to "become round of its own volition" by virtue of the physical process of hydrostatic equilibrium, which happens in objects with a minimum of about half the mass of Pluto, doesn't deserve to be called a planet. Anything in between would be some sort of planet—although if it was orbiting another planet rather than a star, it could be called a planetary body.

Clear. Even persuasive. Yet, despite announcing his opposition to subjecting Pluto's planethood to the democratic process, Stern appealed to a democratic invocation of sorts—the authority of public perception:

I guess the best sort of a test is the test that my favorite fifth grader, my daughter Sarah, suggested. It's the "duh" test. . . . Like the Supreme Court justice on [the definition of] pornography, when it comes to a planet I'm not sure I can give you an exact definition, but I know it when I see it. By the same token, give a fifth grader a picture of Pluto and ask him if it's a planet, and you get back: "Duh."

Marsden, bearing encyclopedic knowledge of the cosmic catalog and willing to agree with everyone onstage, grouped Pluto with its neighbors in the trans-Neptunian Kuiper belt but was perfectly content to embrace dual status—major planet (one of the nine) and minor planet (asteroid)—in the same way as certain small bodies are classed as both asteroids and comets. Once upon a time his hope had been that Pluto would become officially known as number 10,000 in the list of minor planets, in which case, rather than being "a piddly little thing between Mars and Jupiter," it could have been accorded the "nice, unargumentative name" Myriostos, which is Greek for 10,000. Having lost that fight, if a fight it was, he was now amenable to all points of view, provided that the large-enough-to-be-round asteroid Ceres would be treated the same way as Pluto.

Next came A'Hearn, who, like Marsden, was willing to be a dual classifier. He stated his case with utmost and characteristic precision:

Why do we care about classifying Pluto as a planet or as a minor planet, or as anything else for that matter? Why do we do classifications at all in astronomy, or in any other science for that matter? Why do we bother separating humans from chimpanzees?

The reasons we do the classifications is to try to find patterns that will help us to understand how things work or how they came to be. So the way we classify Pluto should be something which helps us to understand how it works or how it came to be, and if what you want to understand is how the interiors of solid bodies work, then you should probably be thinking of Pluto as a planet. If, on the other

> *hand, you want to know how things got to where they are in the solar system, there is no question that Pluto got to where it is in exactly the same way as a large fraction of the other trans-Neptunian objects. . . . So if that's the question you're interested in, you absolutely have to classify Pluto as a trans-Neptunian object. Now, this basically means that you should have dual classification.*

But A'Hearn had a surprise up his sleeve. Classifying Pluto as a trans-Neptunian object would place it in the region that produces comets and would subject it to the IAU's distinctions between comets and asteroids and between comets and minor planets. In other words, if you can see fuzz around an object, it must have an atmosphere, and like icy comets, icy Pluto would have an atmosphere only during perihelion, the few years when its orbit takes it closest to (though still very far from) the Sun's warmth. So, he concluded, "I think it's clear we can come out in favor of comet Tombaugh." The audience loved it.

Last in line was Levy. Invoking the heroism and devotion of Clyde Tombaugh *and* the emotional attachment to Pluto felt by children *and* the cosmic discovery stories recounted to him by his father at the dinner table *and* the meanness of expert taxonomists who decided that the galumphing brontosaurus of his childhood was actually an apatosaurus and that the beautiful Baltimore oriole was actually a northern oriole, he came down squarely on the side of Pluto's planethood:

> *Science, to me, is not just for scientists. Science, to me, is for everyone; it's for us. It's for the children at the Clyde Tombaugh Elementary School. It is for the young people in this audience, who have a better way than we do to look at an object and say, "That's a planet." "That's a brontosaurus."*
>
> *But most important, when we go out under the night sky, and we look up at the stars, we don't see them as being something incredibly complicated, but instead we see them as something beautiful and*

> *simple. . . . Let's send a [spacecraft] out to Pluto. If it gets there, and if it clearly takes a picture that shows that there is a dog instead of a planet, then we can have this debate again, and then we can decide that Pluto is not a planet, it is a dog or a brontosaurus. But until then, please, let's all enjoy the night sky and leave Pluto alone.*

That evening was the first time that we at the Hayden Planetarium—and certainly the people in the audience, or perhaps anybody anywhere—listened to a sustained encounter on the status of Pluto based primarily on the science, but also on culture. And the panelists were divided: one for uncompromising iceballhood, two for dual status, two for planethood. In retrospect, what started as a homework exercise to assist our design of the planet exhibits in the Hall of the Universe was actually a watershed event.

After the speakers' opening remarks, I ran a mental applause meter to take the temperature of the room: Who would be perfectly happy to kick Pluto out of the planet club? Weak applause. Who favors planet status? Modest applause plus a scattering of loud whoops. Yet by the end of the evening, everyone from the Hayden Planetarium connected with the Pluto exhibit design had come to believe that Pluto needn't retain any kind of status at all, except for reasons of nostalgia. And judging by the crowd's laughter and applause as the debate progressed, a majority of them became convinced as well.

Monday, May 24, 1999. The night Pluto fell from grace.

The time quickly arrived for us to design the planet exhibits. We didn't have the power or the authority (or the interest) to declare that the solar system has only eight planets, but that didn't mean we couldn't invent innovative ways to treat the subject. That's when we decided to present the contents of the solar system as families of objects with similar properties, rather than as an enumeration of orbs to be memorized—a trend that was already being seen in textbooks of the day.

One of the "families" is simply our star, the Sun, because it's so much more massive than everything else combined. We then have the terrestrial planets: Mercury, Venus, Earth, and Mars. All of them have more in common with one another than any one of them has with anything else in the solar system. They're small, they're rocky, they're dense, they're near the Sun. Beyond the terrestrial planets, we have the asteroid belt, made up of hundreds of thousands of craggy chunks of rock and metal—debris that never became part of a planet, as well as the fragmented remains of planetesimals that formed but were subsequently shattered. Then come the so-called Jovian planets, the gas giants: Jupiter, Saturn, Uranus, and Neptune. As with the terrestrials, the Jovians have more in common with each other than any one of them has in common with anything else in the solar system. They're big, they're bulbous, they're low density, they're ringed, they're moon-rich, and they're in sequence. Beyond them we have the Kuiper belt of comets, whose orbits all lie more or less in a plane, and then far beyond the Kuiper belt we have a swarm of comets whose orbits go every which way, called the Oort cloud.

Where does Pluto fit? The Kuiper belt. End of story.

We saw no value in counting planets—or counting anything. That exercise to us seemed pedagogically and scientifically vacuous. In an equally unenlightening exercise, consider the answer to "How many countries are there in the world?" There are 192, but that depends on how you define country.[22] The number is as high as 245 if you include places that see themselves as countries, such as Palestine or the Turkish Republic of Northern Cyprus, but are not internationally recognized as such. Want instead to use the official United Nations list? Sounds like a good idea, but that would mean Switzerland was not a country until 2002, when it was finally admitted as a member state. Odd, because

22. CIA *Factbook*; http://www.cia.gov/library/publications/the-world-factbook.

the Swiss city of Geneva is where you could always find one of the four world offices of the United Nations itself, as well as the original location for the League of Nations.

One could also list all the countries alphabetically from somebody's compilation and check off their individual characteristics. But why not simply group countries by "family" resemblances, using data and demographics that tell you something useful, such as region, population size, per capita income, temperature range, life expectancy, or proportion of arable land. Divisions such as these, taken in turn or together, enable you to compare and contrast countries in meaningful ways.

The Rose Center's Cullman Hall of the Universe, named for New York philanthropists Dorothy and Lewis Cullman, is split in four principal sectors: the Planets Zone, the Stars Zone, the Galaxies Zone, and the Universe Zone. In a day gone by, each wall panel in the Planets Zone might have been devoted to a single planet: Mercury and its properties, then Venus and its properties, continuing out to Pluto and its properties. Nine panels. And that would be that.

We did something different.

We looked across the solar system and asked ourselves what physical features about planets and other objects could be taken together and discussed as common properties or phenomena, allowing us to compare and contrast families of objects in whatever way those families would naturally delineate. One such feature is storms; wherever you have a thick, rich atmosphere on a rotating object, you have storms. Another feature is rings. Yet another is magnetic fields. So we took nontraditional cuts through the data of our solar system and presented them among these panels. Pluto was displayed among other Kuiper belt objects, but we neither counted these objects nor made a list of who is or is not a planet.

We knew that no matter how the Pluto debate would ultimately resolve, our familial treatment of the solar system was pedagogically and scientifically sensible—a kind of intellectual high road that sidesteps nomenclature altogether.

The Rose Center for Earth and Space also contains a 400-foot, square walkway that we call the Scales of the Universe. Not to be confused with the famous exhibit on how much you would weigh while standing on various cosmic objects, the Scales of the Universe consists of multiple vistas along a path that surrounds the giant Hayden Sphere. The sphere houses not only the Space Theater but also, in its belly, a walk-through Big Bang experience—a separate venue where we re-create the first few moments of the universe. We further exploit the outside of the sphere as an exhibit element, allowing us to compare things of greatly varied size. With models mounted on the railing of the walkway and suspended from the ceiling, the exhibit shifts by a power of ten in scale for every few yards you walk. At the beginning, the sphere represents the entire universe, and the model on the railing represents the Local Supercluster of galaxies. Take a few steps, and the sphere now represents the extent of the Local Supercluster, while the next model you encounter represents the Milky Way galaxy. Take a few more steps, and the sphere represents the Milky Way, and the model on the railing represents a star cluster. Keep walking, and the scale of comparison keeps dropping and dropping and dropping until you enter the nucleus of the hydrogen atom.

About halfway along this walkway, you reach a vista where the sphere represents the Sun, while models on the railing represent the terrestrial planets, ranging from the size of your fist for Mercury to the size of a cantaloupe for Venus and Earth—all in correct relative size to each other. Suspended near the Sun are scale models of the Jovian, gas giant planets, which are much too large and glorious to mount on the railing.

It's an exhibit on the relative sizes of things. In this picture-perfect spot the visitor is comparing the size of the Sun with that of the terrestrials and the Jovians in one glance. We had no compulsion to include Pluto. Why? We don't show comets. We don't show asteroids. We don't show the seven

gure 4.6. *Models representing terrestrial planets, seen from Scales of the Universe walkway the Rose Center for Earth and ace. With the Sun represented the Hayden Sphere (see Figure), models of the terrestrial anets, in correct relative size, mounted on the railing. On s scale, Mercury (left) is a little ger than a baseball; Earth and nus, soccer balls; and Mars ght), a bocci ball. Pluto, not ing a terrestrial planet, does not pear among them. (Earth serves the principal size referent to the n and so remains unpainted, ong with other fiducial models of exhibit.)*

gure 4.7. *View from the Scales the Universe inside the Rose nter for Earth and Space. The ayden Sphere, representing the n from this vista, is juxtaposed th the gas giant (Jovian) planets piter, Saturn, Uranus, and ptune, each suspended from ceiling. Pluto, not being a ian planet, does not appear ong them. For any given scale, models are constructed and splayed in correct relative size to ch other.*

moons in the solar system that are bigger than Pluto. We were making a simple comparison of how two families of objects in the solar system contrast with the Sun in size.

If you want to learn about the form and structure and contents of the solar system, then visit the Planet Zone in the Hall of the Universe. That's where you will find Pluto, in a transparency, by the way, huddled with its fellow members of the Kuiper belt.

The new Rose Center for Earth and Space opened to the public on Saturday morning, February 19, 2000. We were not unmindful of the potential for controversy over how we had organized the planet exhibits. But at no time during the hundreds of radio, print, and television interviews I gave—for domestic as well as international media—did our treatment of Pluto come up. I didn't volunteer it either. Actually, the *New York Post* and one or two other regional papers, in their preview of the facility, noted the absence of Pluto in our Scales of the Universe, but none of them made a big deal of it. The coast was clear. No media controversy.

But that was just the calm before the storm.

Then one day, a reporter for the *New York Times*, on his own time, decides to visit the Rose Center and just have a look around. On the Scales of the Universe walkway, the reporter overhears a child asking his mother, "Mommy, where's Pluto?" She replies with unjustified confidence, "Check again, you're not looking hard enough."

The child repeats, "Mommy, where's Pluto?" Of course, neither of them can find Pluto because Pluto isn't there. Meanwhile, the eavesdropping reporter is sure he's got a story. So he calls the paper, and they put Kenneth Chang on the case. Chang, an eager, smart, young science reporter, does some investigating of his own and files a story for the *Times*.

On January 22, 2001, nearly a year after opening day for the Rose Center, a full news day had passed since George W. Bush was inaugurated as the 43rd president of the United States. The election was controversial. The dimpled chads from Florida's paper ballots were still flapping in the breeze. One might expect that day's front page of the *New York Times* to be filled with stories from Washington and elsewhere, chronicling reactions to the new American president.

Page 1 was indeed headlined "On First Day, Bush Settles Into a Refitted Oval Office—He Greets Public After Touring New Home." But that article also shares space with other important stories, including one on U.S. intelligence estimates concerning three alleged Iraqi weapons factories, another on the pope's newly appointed cardinals, and still another that reports on California's race to build power plants so they may avert their perennial summertime brownouts.

And there it was. Page 1. Kenneth Chang's article on the Rose Center. Appearing in 55-point type was the headline that would disrupt my life for years to come:

PLUTO'S NOT A PLANET? ONLY IN NEW YORK

The article continued across four columns of carryover, including a photo, and a diagram in Section B.

Chang opened by retelling the frustrations of Atlanta visitor Pamela Curtis, who had to exhume from memory, then recite out loud, the time-honored planet mnemonic "My Very Educated Mother . . . " to establish that Pluto was, in fact, missing from the display of orbs. Then Chang went for our jugular, portraying our approach to Pluto as both renegade and suspect:[23]

> *Quietly, and apparently uniquely among major scientific institutions, the American Museum of Natural History cast Pluto out of the pantheon of planets when it opened the Rose Center last February.*

23. Kenneth Chang, *New York Times*, January 22, 2001, pp. A1, B4.

> *Nowhere does the center describe Pluto as a planet, but nowhere do its exhibits declare "Pluto is not a planet," either. . . .*
>
> *Still, the move is surprising, because the museum appears to have unilaterally demoted Pluto, reassigning it as one of more than 300 icy bodies orbiting beyond Neptune, in a region called the Kuiper Belt.*

Then came the quotes from the experts. This from planetary scientist Richard Binzel, of MIT (see Figure 3.10): "They went too far in demoting Pluto, way beyond what the mainstream astronomers think." And this from Alan Stern, who we met on the museum's Pluto panel: "They are a minority viewpoint. . . . It's absurd. The astronomical community has settled this issue. There is no issue."

But if you followed the article to page B4, you learned that astronomers had been reconsidering Pluto for years:

> *The International Astronomical Union, the pre-eminent society of astronomers, still calls Pluto a planet, one of nine of the solar system. Even a proposal in 1999 to list Pluto as both a planet and a member of the Kuiper Belt drew fierce protest from people who felt that the additional "minor planet" designation would diminish Pluto's stature. . . .*
>
> *But even some astronomers defending Pluto admit that were it discovered today, it might not be awarded planethood, because it is so small—only about 1,400 miles wide—and so different from the other planets. . . .*
>
> *As a planet, Pluto has always been an oddball. Its composition is like a comet's. Its elliptical orbit is tilted 17 degrees from the orbits of the other planets. . . .*
>
> *But Pluto continued to be called a planet, because there was nothing else to call it. Then, in 1992, astronomers found the first Kuiper Belt object. Now they have found hundreds of additional*

chunks of rock and ice beyond Neptune, including about 70 that share orbits similar to Pluto's, the so-called Plutinos.

Buttressing the case was a diagram titled "To Be, or Not to Be, a Planet," showing Pluto as a midway point between the planets Earth and Mercury on one side, and the not-planets Ceres and 2000 EB173 on the other; its label stating that "Pluto is bigger than minor planets and has an atmosphere," and yet "Pluto has an unusual orbit and is made largely of ice."

Chang quoted me in several paragraphs, where I defended our treatment of Pluto. And in the final paragraph, I got to reprise the concluding words from my "Pluto's Honor" essay with the comment that Pluto would surely be happier as king of the Kuiper belt than as the puniest planet. One of the later remarks came from a scientist at the Denver Museum of Nature and Science, which was building a new space sciences center but would continue to display Pluto as one of nine:

"We're sticking with Pluto," said Dr. Laura Danly, curator of space sciences at the Denver museum. "We like Pluto as a planet."

But, she also said, "I think there is no right or wrong on this issue. It's a moving target right now, no pun intended, what is and is not a planet."

People don't always read articles to the end. Laura Danly's candor that Pluto's classification was a moving target came in the last column. We would later attract Danly from Colorado to become our museum's director of astrophysics education before she would move once more to become curator of education at the newly renovated Griffith Observatory and Planetarium in Los Angeles.

In the end, it was the *Times* headline writer who got the last word. "Pluto's Not a Planet? Only in New York" became the takeaway, rather than the more accurate (unwritten) title "Pluto's Not a Planet? A Growing Number of Professionals Agree."

So on January 22, 2001, beginning at about 7 a.m., my phone started ringing. My voice mail filled (I never knew before that day the voice mail capacity for our phone system). My e-mail in-box overflowed. And my life would never again be the same.

It's always a little scary when the person who hired you calls you up and asks, "What have you done?!" In my case, the fellow on the phone was Michael Novacek, an accomplished paleontologist and the museum's provost. Keep in mind that I was relatively new at the institution—a youngish upstart in charge of the science content for $230 million worth of the museum's money.

Of course, the museum, a research institution as well as a place where you find exhibits, is right to be concerned about its scientific integrity. So when it gets dragged through the mud by a page 1 news story about its treatment of a scientific issue, the people on top want to know why. And so, perhaps out of concern that I had strong-armed my own personal view into an institutional posture, Novacek asked whether I had downgraded Pluto on my own or had built some kind of consensus behind the decision. My answer, of course, was that all internal and external members of my scientific advisory committee had reviewed the matter, and the exhibit treatment represented a consensus.

Regardless, the museum sought a quick second opinion from an authoritative source. So Novacek called Jeremiah P. Ostriker, author of more than 200 scientific papers on astrophysics, winner of the U.S. National Medal of Science, provost of Princeton University and former chair of Princeton's renowned department of astrophysical sciences (where I was, at the time, an adjunct professor), and a recently appointed trustee of the museum itself. What did Ostriker say to Novacek? "Whatever Neil did is okay by me."

I didn't learn of this exchange until years later, when Ostriker casually recounted it as part of another conversation. He had treated the media attention like a non-event, the same way Jane Luu had done in the panel debate. To them, the hoopla wasn't about a scientific question. The organization of the solar system, how the solar system came to be the way it is—those are genuine scientific questions. But the labels you give things—no. You're having an argument over something you generate rather than what is fundamental to the universe. While you're sitting around debating, Pluto and the rest of the universe happily keep doing whatever it is they do, without regard to our urges to classify.

Meanwhile, just weeks after Kenneth Chang's article appeared in the *New York Times*, the same paper published a second article, this one the brainchild of Mark Sykes, a planetary scientist at the University of Arizona's Steward Observatory in Tucson and, at the time, chair of the American Astronomical Society's Division for Planetary Sciences. Well aware of the storm brewing at the Rose Center, Sykes sent an e-mail warning me that the division's executive committee might be drafting a statement rebuking our treatment of Pluto. He also alerted the *Times* that he would be in New York on business and intended to meet with me to discuss the matter—and would the *Times* like to send over a reporter to listen to the conversation? They, of course, agreed.

Sykes came. So did Kenneth Chang, serving as witness and juror, as well as a *Times* photographer. We chatted around my office's brass coffee table, retrofitted from a 4-foot, circular, contemporary engraving that had been on display among the history of science exhibits of the old Hayden Planetarium. It portrays the long-defunct geocentric model of the universe, complete with Earth in the center and planetary epicycles looping around it. Not to miss a single word, Chang recorded the entire conversation on tape.

The resulting article appeared with the headline: "Icy Pluto's Fall From the Planetary Ranks: A Conversation," accompanied by a short piece about the prospects for a mission to Pluto. There's also a photo of Sykes attempting to choke me alongside the giant sphere, with the gas

Figure 4.8. Mark Sykes (left) and the author in adversarial, but playful embrace on the Scales of the Universe walkway of the Rose Center for Earth and Space. This photo appeared in the New York Times, *accompanying the transcript of our minidebate. Sykes, a planetary scientist and, at the time, chair of the Planetary Sciences Division of the American Astronomical Society, threatened to have his division draft a public statement rebuking the treatment of Pluto in our exhibits. I threatened to toss him over my left shoulder into the Hall of the Universe below.*

giants hanging in the background. The caption reads, "Dr. Mark Sykes, left, challenges Dr. Neil deGrasse Tyson to explain the treatment of Pluto in the planet display at the Hayden Planetarium."

The published conversation reads like a raw transcription, and as we see from this excerpt, Sykes's views on the matter are unyielding and unambiguous:[24]

> **DR. SYKES** *The consensus exists. Unanimity may not, but I think consensus does, and the consensus is that people feel Pluto should not—it's fine to call it a Kuiper Belt object—but we should not*

24. Kenneth Chang, *New York Times*, February 13, 2001, p. F2.

remove its designation as a planet. People are thinking not families, not groups, not cousins. They're thinking planets. . . .

When people come in, they are expecting to see what astronomers think. What you've got up here is not what astronomers think. . . .

DR. TYSON *It's what some astronomers think. . . .*

DR. SYKES *Some astronomers that I can think of, that I can put on one hand. . . . I would say were Pluto discovered today and known to have a moon and an atmosphere, I think that it would be designated a planet and not just given a minor planet designation. . . .*

It's got nitrogen ice caps. It's got seasons. It's got a moon. It's got an atmosphere. It's got a whole suite of properties which distinguishes it from what we know about any other Kuiper Belt object, and just to blithely say, Well, we're just not going to tell you about this and we're just going to lump it in with these other guys, is, from an educational standpoint, irresponsible. . . .

If Pluto were 10 times its size, how would you treat it?

DR. TYSON *I think if it were still ice, we'd still say, orbiting with the icy objects.*

DR. SYKES *Pluto is thought of as a planet. So why not icy planets. Pluto.*

DR. TYSON *With a class of one?*

DR. SYKES *Class of one. Sure. Why not?*

I would later learn that in addition to his PhD in planetary science from the University of Arizona, Mark Sykes also wields a law degree from the same institution and was admitted to the Arizona bar. So this, from my point of view, accounted for at least some of his urge to argue.

The media was unrelenting. For every story, every account, and every analysis, the story grew in the hearts and minds of the public, with the Rose Center continually landing in the middle of the credit or the blame.

One mailing I received in May 2001 contained 100 pages of typed essays from Mr. James Dixon's ninth-grade honors earth science class at Silver Lake Regional High School, in Kingston, Massachusetts. Invoking my published writings on Pluto as well as media accounts and other sources, each student pleaded a case. The votes were evenly split among Plutophiles and Pluto demoters—a rare result this early in the debate. Regardless, here was a resourceful educator turning a science controversy into a teaching opportunity—the pedagogical equivalent of using lemons to make lemonade. I applaud his effort as well as that of other teachers who organized their curriculum to include topical discussions of Pluto's status.

This was not the first time Mr. Dixon had corresponded with me. Two years earlier, before the Rose Center had opened to the public, he and his class had sent in their first stack of letters, in response to my "Pluto's Honor" essay for *Natural History* magazine. At the time, I didn't know I had a Pluto pen pal in the making.

Not much time would pass before people started to see the humor in it all. A February 16, 2001, op-ed for the *New York Times*, written by Eric Metaxas, a writer for Veggie Tales Children's Videos, could not resist a tally of abominable headlines to come, but only if our treatment of Pluto started a cultural trend:[25]

TROPICS OF CAPRICORN AND CANCER TO BE ABOLISHED

TOMATO UNEQUIVOCALLY DECLARED A VEGETABLE

GREAT LAKES TO BECOME FIRST FRESHWATER OCEAN

BEIGE TAUPE AND MAUVE LOSE COLOR STATUS

60S STARTED IN 1963

25. Eric Metaxas Op-Ed, *New York Times*, February 16, 2001, p. A19.

Texas Declares Itself a Subcontinent

European Maps to Drop Lichtenstein

Meter and Yard Shake Hands

Asterisk Removed from Gerald Ford's Presidency

Tab and Zima Lose Beverage Status

Hard to argue with most of them.

Over the next few years, as anticipated by Jewitt and Luu and many other sky watchers, more and more Kuiper belt objects were discovered. Most were icy and traveled in eccentric orbits. Those that most resembled Pluto in orbital parameters came to be called Plutinos. Several of these objects rivaled Pluto's moon Charon, if not Pluto itself, in mass, size, and properties. For planetary scientists, the region of space beyond Neptune became ever more populous and ever more intriguing.

Let's look at diameters. In November 2000, using the Spacewatch 0.9-meter telescope at the University of Arizona in Tucson, Robert S. McMillan spotted an object now officially named 20000 Varuna, estimated to be almost 900 kilometers wide.[26] In a May 2001 digital image obtained by James L. Elliot and Lawrence H. Wasserman using the 4-meter Blanco Telescope at Cerro Tololo, Chile, a team of astronomers from the NASA-funded Deep Ecliptic Survey spotted the similarly sized 28978 Ixion. And in June 2002, using the 48-inch Oschin Telescope at Palomar Observatory in southern California, Caltech astronomers Chad

26. *Spacewatch*; http://spacewatch.lpl.arizona.edu/2000wr106.html.

Trujillo and Michael Brown spotted 50000 Quaoar.[27] With a diameter of some 1,300 kilometers, almost half that of Pluto, Quaoar was the largest thing found in the Kuiper belt since Clyde Tombaugh found Pluto in 1930. Quaoar's discovery was announced to the world on October 7, 2002, at a meeting of the American Astronomical Society, tripping an awareness threshold at the *New York Times*. Eight days later, the newspaper ran an editorial headed "Pluto's Plight":[28]

> *Things are looking bleaker than ever for Pluto, the most disrespected of the nine planets that we learned about in elementary school. . . .*
>
> *Although Pluto's fans hate to admit it, the ninth planet owes its status more to the fact that astronomers expected to find a planet out beyond Neptune than to any intrinsic merit. . . . Last year the Hayden Planetarium caused quite a stir by dropping Pluto from its list of planets.*
>
> *Now, in the latest blow, astronomers reported last week that they have found another dirty ice ball, about half the size of Pluto, that actually behaves more like a planet than Pluto does, with a circular orbit. The newly discovered Quaoar (pronounced KWAH-o-ar) lies among a multitude of small bodies in a region known as the Kuiper Belt. . . .*
>
> *Astronomers predict that they will find up to 10 similar objects in the Kuiper Belt that are as large as or larger than Pluto. So unless we want to add 10 more planets to the elementary-school curriculum, we would be wise to downgrade Pluto to the distant iceball it is.*

Excuse me? Was this the same newspaper whose page 1 headline blamed Hayden for all of Pluto's woes? And what's that about the Hayden

27. By international agreement, trans-Neptunian objects are named after creation deities. Quaoar comes from the Native American Tongva people, who are local to the Los Angeles area that includes Caltech, the home institution of its discovery.
28. Editorial, *New York Times*, October 15, 2002, p. A26.

Figure 4.9. Eight trans-Neptunian objects drawn to scale, including Pluto. As the inventory of objects that orbit the Sun beyond Neptune grows, so, too, does the likelihood of finding objects that rival Pluto in size. Already, Eris checks in larger than Pluto. More will surely follow. For size reference, a segment of Earth appears along the lower edge.

Planetarium causing "quite a stir"? Wasn't it the *New York Times* that broke the story—a year late—and not some kind of media press release on our part? We had been quietly minding our own planetary business for a year before they even took notice. And what's that about "we would be wise to downgrade Pluto"? Who's "we"?

All during the previous year, the *Times* had allied itself with the Pluto-is-a-planet people. But if the *Times* was now (albeit tacitly) acquiescing to the sensibility of our approach—if the paper was finally seeing what we saw the night of our Pluto panel—then we were happy to have them on our side.

A year later, in November 2003, a reddish ball now named 90377 Sedna, with a diameter somewhere around 1,500 kilometers—about three-quarters that of Pluto—was discovered by Kuiper belt hunters Michael Brown, Chad Trujillo, and David Rabinowitz, again using the Oschin Telescope. It was the most distant known body in the solar system, and planetary scientists were rapidly closing in on the Holy Grail—an object out there bigger than Pluto itself.

In fact, it had already been found the previous month. On October 21, 2003, Brown, Trujillo, and Rabinowitz photographed, but only later revealed, the ninth largest solar system body in orbit around the Sun: 136199 Eris, whose diameter falls somewhere between 2,400 and 3,000 kilometers (Pluto's checks in at 2,300 kilometers) and whose mass is 27 percent larger than that of Pluto. Brown and his collaborators announced their discovery along with two other (smaller, less planet-shaking) objects on July 29, 2005.

If Eris were to keep its classification as a planet under IAU rules, as Mike Brown argued should happen, then his object was a shoe-in for planet status, making Brown the second American to discover a planet, after Clyde Tombaugh. But the converse must also be true: If Eris were classified as something other than a planet—how about a dwarf planet?—then Eris, being bigger than Pluto, would drag Pluto down with it. Such was the battlefield drawn by the inevitable discovery of Eris.

Before Eris was formally named by the IAU in September 2006, Mike Brown had unofficially been calling the thing Xena, after the buff, buxom, leather-clad, sword-wielding warrior princess of cable television who spends much of each weekly episode kicking medieval butt. Unfortunately (in my opinion), television mythology is not a valid source of names for cosmic objects. The IAU instead references mythologies from a time somewhat before TV was invented. Mike Brown knew this and had proposed the name Eris, the Greek goddess of discord and strife. For Eris's moon, whose orbit enabled the accurate calculation of Eris's Pluto-beating mass, he proposed Dysnomia, the demon goddess of lawlessness and the daughter of Eris. As you may already know, classical gods led complicated social lives. One of Eris's pastimes was to instill jealousy and envy among men, driving them to battle. At the wedding of Peleus and Thetis, all the gods were invited with the exception of Eris. Angered by her exclusion, she vengefully instigated a quarrel among the goddesses that precipitated the Trojan War.

Brown had indeed done his classical homework and dutifully captured Eris's destabilizing influence on the Pluto problem, causing a war of its own.

Pluto's a Planet protest group.

5

Pluto Divides the Nation

IT'S NOT EASY BEING A PUBLIC ENEMY. NEARLY ALL the paper mail and e-mail that I received following the trouble-making January 22, 2001, *New York Times* article was negative. Schoolchildren and adults alike branded me a thoughtless, heartless Pluto hater. Some, not paying attention to the one-year delay of the *Times* article from opening day of the Rose Center, accused the museum of mounting a publicity stunt to drum up attendance.

I personally replied to nearly every inquiry. Most letters presumed that we, the mean, city slicker New Yorkers, had kicked small, defenseless Pluto out of the solar system and that I was somehow personally trying to disrupt the nine-planet family we had all come to befriend in elementary school.

Of course, what we actually did to Pluto was more subtle than this, which prompted me to compose a "media response" to the barrage of inquiries, attempting to clarify our position compared with what someone would think we did based on the *New York Times*' "Pluto's Not a Planet? Only in New York" headline. The content of this 1,000-word statement (see Appendix E for the complete text) includes exact quotes from our exhibits. In the Planet Zone of the Hall of the Universe, we ask the question, "What is a Planet?" to which we reply:

> *In our solar system, planets are the major bodies orbiting the Sun. Because we cannot yet observe other planetary systems in similar detail, a universal definition of a planet has not emerged. In general, planets are massive enough for their gravity to make them spherical, but small enough to avoid nuclear fusion in their cores.*

This paragraph is hardly controversial. Our exhibit goes on to describe "Our Planetary System":

> *Five classes of objects orbit our Sun. The inner* ***terrestrial*** *planets are separated from the outer* ***gas giant*** *planets by the* ***asteroid belt****. Beyond the outer planets is the* ***Kuiper Belt*** *of comets, a disk of small icy worlds including Pluto. Much more distant, reaching a thousand times farther than Pluto, lies the* ***Oort Cloud*** *of comets.*

Another innocuous paragraph. But the persistent (negative) attention received by the museum forced us to temper the storm. After I review the overall layout of the Rose Center, I describe in detail the "offending" part of our Scales of the Universe exhibit but remain firm about our intent:

> *About midway in the journey [along the Scales of the Universe] you come upon the size scale where the sphere represents the Sun. On that scale, hanging from the ceiling, are the Jovian planets (the most highly photographed spot in the facility) while a set of four small orbs are also on view, attached to the railing. These are the terrestrial planets. No other members of the solar system are represented here. This entire exhibit is about size, and not much else.*

I then address the issue head-on:

> *But the absence of Pluto (even though the exhibit clearly states that it's the Jovian and terrestrial planets that are represented) has led about ten percent of our visitors to wonder where it is.*

Ten percent of the public bewildered is a large enough fraction to concern us as educators. The release continues:

> *In the interest of sound pedagogy we have decided to . . . add a sign at the right spot on the size scales exhibit that simply asks "Where's Pluto?" and gives some attention to why it was not included among the models.*

Shortly thereafter, we wrote, designed, manufactured, and bolted a "Where's Pluto?" plaque to our Scales of the Universe walkway, adjacent to and visible from the rail-mounted models of Mercury, Venus, Earth, and Mars. No longer would people ask Pluto's whereabouts among our exhibits, but these measures did nothing to stave off the fulmination that would follow.

We released the statement to general audiences on February 2, 2001, but it was specifically intended for the widely read, UK-based Internet chat group called Cambridge Conference Network (CCNet), moderated by social anthropologist Benny J. Peiser, of Liverpool John Moores University. The primary interest of the network was open discussion of asteroids, comets, and their risk to life on Earth, but many other newsy subjects also found their way to these pages.

On January 29, 2001, Peiser posted articles from the Associated Press (AP) and *Boston Globe* that had been spawned by the original *New York Times* story on Pluto a week before.[29] The AP article contained a quote from me:

> *There is no scientific insight to be gained by counting planets. Eight or nine, the numbers don't matter.*

This was followed by a quote from David Levy, amateur astronomer extraordinaire (who we met in the last chapter at the museum's 1999 panel on Pluto's status). Levy drew first blood with the barb:

> *Tyson is so far off base with Pluto, it's like he's in a different universe.*

Always remember that when an astronomer accuses you of being in a "different universe," it carries extra meaning.

Sustained, verbal altercations followed immediately in the chat group. University of Hawaii's David Jewitt, codiscoverer of the Kuiper belt of

29. All quotes from CCNet used with permission from Benny J. Peiser.

comets (along with Jane Luu, who we also met in our discussion of the Pluto panel), fully endorsed our exhibit treatment of Pluto:

> *They've done exactly the right thing. It's an emotional question. People just don't like the idea that you can change the number of planets. It's inevitable that other museums will come around, though. The Rose center is just slightly ahead of its time.*

Leonard David, journalist from Space.com, quoted space scientist Kevin Zahnle as saying:

> *Pluto is a true-blue American planet, discovered by an American for America.*

I later learned from a colleague that Zahnle, of NASA's Ames Research Center, in Moffett Field, California, is only capable of such a statement in jest, but others who did not know this took the jingoistic comment seriously. Joshua Kitchener, publisher of a Web-based asteroid-tracking magazine, replied immediately:

> *Such romanticism has no place in science, a system which must never cease trying to determine the objective truth, a truth free from human prejudice and emotion. Neither does nationalism.*

Want to piss off the astronomers?[30] Call them astrologers. Space scientist Wendell Mendell, of NASA's Johnson Space Center in Houston, did just that:

30. Professional astronomers today, including planetary astronomers, are basically astrophysicists. We all have extensive training in mathematics and physics and seek to understand the operations of nature as expressed by the laws of physics. So in modern times, the terms *astronomer* and *astrophysicist* are interchangeable, but with *astronomer* carrying a historical link to the days when all we could do was tell you where something was in the sky and what it looked like through a telescope.

> *I confess I am disappointed in the learned community that joins with astrologers in holding onto an outdated classification scheme.*

I was particularly charmed by the attempt of some to make peace among the combatants. Dale Cruikshank, president of the IAU's commission for the physical study of the planets and satellites, took both sides of the argument:

> *My personal view is that Pluto should probably have dual citizenship, in that its planetary status ought to be maintained partly for historical reasons and partly for its physical characteristics. But it seems clear that it is also "object one" in what we now recognize as a large class of Kuiper Belt objects.*

David Levy, with perennial concern for the relationship between scientists and the public, commented:

> *On the whole, I do not agree with the dual status because it complicates matters too much in the public perception.*

Levy's comment here is entirely consistent with his concern-for-the-public testimony at the museum's Pluto panel. A day earlier, Levy had posted a claim that left you wondering whether he was a personal friend of Pluto:

> *I believe that until we land on Pluto and find incontrovertible evidence that that world does not wish to be called a planet, that we should leave things as they are.*

My professional research interests relate primarily to stars and galaxies. Knowing this, geologist Jeff Moore, from NASA Ames, chose to attack my absence of professional association with the solar system:

> *First of all, it's rather amazing that Tyson, an astrophysicist, would even venture into such waters. I feel, as a planetary geologist, equally qualified to demote the Magellanic Clouds*[31] *to glorified star clusters as opposed to small galaxies. So in that spirit, I think he's full of baloney.*

I have always enjoyed seeing the word *baloney* used in a sentence.

Returning to the dialogue, Joshua Kitchener could not resist a historical analogy to the proceedings:

> *It's not too hard to imagine the same type of people back in Galileo's age, saying, "I've been taught that the Earth is the center of the universe since I was a child. Why change it? I like things the way they are."*

Mark Kidger, of the Institute for Astrophysics in the Canary Islands, followed with a perceptive comment about the trans-Neptunian objects (TNOs) of the Kuiper belt:

> *Had other TNOs been discovered in 1935 and not in 1992, it is quite possible that we would not be having this debate now.*

These views, and many more, were all posted to CCNet within a single 24-hour period. Benny Peiser would now solicit the open letter of rebuttal from me, with a polite appeal that concealed much of the emotion that the raging conversation had engendered:

> *I regret that you have received a considerable amount of flak and criticism for your pioneering decision, and would like to congratulate you for your courage. As a moderator of CCNet, I have tried to*

31. The two nearest systems to our Milky Way galaxy are the large and small Magellanic Clouds. Visible principally from the Southern Hemisphere, the explorer Magellan, during his round-the-world journey, thought they were clouds. But telescopes later revealed them to be "dwarf" satellite galaxies in orbit around the Milky Way.

keep this whole debate in the domain of fact and evidence, rejecting right from the start any attempts to intimidate those who suggested changes to the status of Pluto.

Please let me know whether you would be happy to write a little essay-type comment for CCNet and its many readers. I would very much [like] to hear from you.

On February 14, 2001, Benny Peiser decided to post the entire *New York Times* article that contained the conversation between me and Mark Sykes, chair of the American Astronomical Society's Division for Planetary Sciences. A week earlier, in a kind of preamble to that posting, Mark Sykes may have been concerned about whether the *New York Times* article adequately captured his views. So he submitted a 900-word letter to CCNet bluntly recounting his disapproval with what he saw among our exhibits at the Rose Center. He made his point of view quite clear:

. . . the issue at the Hayden is more one of poor pedagogy than a clarion call for controversy.

The exhibit on planets at best confuses those who look closely enough to catch the inconsistencies. . . .

. . . scientific and pedagogical integrity would require that they prominently notify the public that they are taking an advocacy position to remove Pluto's planetary status and acknowledge that at present the IAU officially designates Pluto as a planet.

That letter did not go unheeded. University of Memphis astrophysicist Gerrit Verschuur struck back as though it was he, and not the Rose Center, who had been attacked personally:

I was shocked to read the following from Mark Sykes . . . "When designing an exhibition, one needs to understand and take into consideration the expectations of the viewer. Given an opportunity, the viewer will see what they expect to see." Surely an exhibition that does just that contributes nothing. If viewers only see what they expect to see they might as well stay home. Does the Sykes philosophy mean when designing an exhibit about UFOs in which one hopes to educate that one must give the viewers what they expect to see, which is a load of nonsense about aliens flying between the stars? . . . Surely the point of an exhibit in science is to inform and educate and not just to feed prejudices and expectations.

Verschuur goes on to make an insightful pedagogical observation about teaching Astro 101:

I feel sure most of us who have taught astronomy have felt troubled when we reach the chapter that shows Pluto looking at us from the end of chapters on the gas giants, or even worse, lurking among the terrestrial planets.

Verschuur, author of five popular books on astrophysics, continues with a second insightful observation, referencing the ruffled relations between pure research scientists and scientists who also choose to bring the frontier of science to the public:

Is the problem perhaps that the Pluto controversy has been stirred up by a planetarium, given that many professional astronomers are still inherently prejudiced against anyone who deigns to dedicate their time to the popularization of astronomy?

The hidden folly of it all was succinctly captured by Sonoma State University astronomer Phil Plait:

At the heart of the debate is our very definition of the word "planet." Currently, there isn't one. The International Astronomical Union (IAU), a worldwide body of astronomers, is the official keeper of names. It has no strict definition of a planet, but has decreed that there are nine major planets, including Pluto. This, however, is not very satisfying. If the IAU doesn't really know what a planet is, how can it know there are nine?

Mark Sykes was not alone in his brazen ways. Many of my colleagues felt comfortable telling me directly, via e-mail, what their opinions were regarding our exhibit treatment of Pluto. They wasted no time, most arriving at my in-box within days of the story breaking in the *New York Times* on January 22, 2001, and with others trickling in over the years that followed.

Robert L. Staehle, of NASA's Jet Propulsion Laboratories, in Pasadena, California, sure that we were guilty of a simple oversight, candidly wrote:

What gives? Did someone there have a memory lapse? What will it take to get Pluto back up there where it belongs?

Staehle would later follow with an observation of nature that we can all agree with:

In the end, neither Pluto nor anything else in the outer Solar System cares in the slightest what anybody on Earth labels it. These bodies exist, and hold what they have to reveal about our collective natural history, totally without regard to any label assigned by an august body of scientists, or any other living thing that we know of.

Michael A'Hearn, of the University of Maryland, who was kind enough to participate in our Pluto panel two years earlier, shared a perceptive pedagogical observation:

> *I am struck by the fact that the people who seem most wedded to keeping Pluto a planet are systematically the people who do not have to teach the subject (either to students or the public) on a day to day basis.*

Timothy Ferris, the best-selling science writer, was quick to offer support, combined with a prescient statement of the long-term outcome of our decision:

> *Unhindered by any prejudices on the subject other than a personal liking for Clyde Tombaugh, a good man whose advice was central in my design of Rocky Hill Observatory, I reviewed the case and concluded that, hell, Pluto is not a planet. So I think you guys did the right thing, and will in the long run be viewed as having led the bird when your only alternative was to miss.*

Long before Mark Sykes had come to New York City to get in my face and argue about Pluto, he sent an e-mail that was entirely consistent with his long-standing posture on the subject, outraged and full of energy:

> *I recall the "issue" starting out as a joke by Brian Marsden at a party in the late 80s. It seems the joke is on your institution. Unfortunately, it is the public who loses out. Minority opinions can be a wonderful starting point for an illuminating discussion, but to the extent your exhibit is meant to be educational, it should be identified as such and the argument engaged. Otherwise, you make a misrepresentation, in your silence, of the view commonly held by planetary scientists.*

My Pluto in-box would not be complete without a quip from Alan Stern, sent years after the exhibit dustup. In a postscript to a letter on another subject, Stern was characteristically terse and firm, punctuated with a winking smiley face at the end:

It's a planet, man. You gotta get over this. ;-)

Bill Nye the Science Guy®, (seen in Figure 3.11), a consummate educator and hobbyist of linguistics, opined with a tutorial on much-needed nomenclature in the field of planetary science:

The great thing about this debate is that it has gotten people everywhere thinking about planets and our place among them. It's remarkable. The whole world it seems is full [of] people puzzling Pluto.

Words include more than they leave out. Words never say all there is to say about anything. So, I advocate some adjectives. I favor Pluto, Xena (or whatever its ultimate sobriquet), Sedna, and others being called generally "planets." Then we'd have teaching opportunities with the adjectives or descriptors:

"Main Plane planets" (those in the plane of the ecliptic)

"Ice dwarf planets" or even "Plutonian" planets (spherical icy worlds akin to their namesake Pluto)

The expression Main Plane has assonance and is thereby easy to say and remember.

If all that was not enough, a brief lesson in Latin followed:

There are apparently untold "Plutonian," or "Ice Dwarf," planets beyond Neptune. These would or should be described as "ultra-Neptunian" planets. Nota bene, I feel strongly that my Latin teachers would suffer greatly to know that certain of my astronomical colleagues use the Latin "trans" to mean "beyond." Sigh. "Trans" means "across." There are occasional usages of traneo for "go past." But to my ear, it's not the same. For "beyond," Romans use "ultra." I hope the nomenclature committee(s) can come around on this one.

There aren't many of them in the world, but Geoff Marcy, of the University of California at Berkeley, is one: a hunter of exoplanets.[32] In reference to the views of Mark Sykes and the need to not rock the boat, Marcy had this to say:

He feels that mere science museums should communicate only the IAU party line. That's not how I read the constitution, and it's not how I read productive scientific discourse. Discussion of the issues should prevail, especially when the observables lean clearly one way. That the IAU has political agendas and recalcitrant members is no reason for us to shade the truth about Pluto.

Don Brownlee, planetary scientist at the University of Washington, was simple and direct with an opposing view:

Demoting Pluto to just another KBO is revisionist science and a cheap shot at history.

Of course, one might alternatively view revisions in science as a good thing—a sign of progress and discovery.

Wesley Huntress, former associate administrator of space science at NASA, wrote from the Carnegie Institute of Washington, where he served as director of its geophysical laboratory. After a brief admonishment,

32. Exoplanets are planets that orbit stars other than the Sun (compare with exobiology). They're sometimes clumsily called extrasolar planets.

questioning whether the Rose Center has fallen wayward of scientific consensus, he reasons himself back to a position not fundamentally different from our own:

> *The Science Citadel of the Capital of the World should not confuse the public. Given the continuing discovery of Kuiper Belt Objects, and in particular, objects larger than Pluto and also having their own satellites, we need a new map of the "world" of the solar system as we explore its wider seas. . . . Our solar system has two belts of multiple small objects, the first between Mars and Jupiter containing rocky bodies, and the second beyond Neptune stretching far into interstellar space to the Oort Cloud that contains icy objects. . . . So we have an asteroid belt and a comet belt. The objects within them can be called minor planets if they are self-gravitating and therefore round; rock dwarfs in one case and ice dwarfs in the other. Otherwise they are asteroids or comets. Ceres is a rock-dwarf, Pluto is an ice-dwarf, and there are eight planets in the solar system.*

Huntress ends by reflecting, as only a wise, empty-nest parent would—

> *Sometimes you have to let your children go.*

Of course scientists were not the only ones whose opinions came to my in-box. Before there was the *New York Times*, before there was CCNet, before there was Mark Sykes, before the New Horizons mission was sent to Pluto, there was Will Galmot, the first person to notice that Pluto was missing from our displays and to write us about it. This astute visitor to the museum did so ten months before the Rose Center's Pluto story broke in the press. Mr. Galmot was apparently paying closer attention than everyone else during our first month of operation and had carefully researched the problem. Reproduced in Figure 5.1, Mr. Galmot's terse correspondence was clear and to the point. And in case we were unsure of what Pluto looked like, Will used the artistic tools available to him and supplied a detailed image for our exhibit professionals to use.

By mid-2001, I was receiving organized packs of student letters every several weeks, sent by zealous teachers eager to tell me how their class voted on the matter. In June 2001, Miss Fedi's fourth graders from Dean La Mar Allen Elementary School in Las Vegas, Nevada, voted 90 percent to 10 percent in favor of retaining Pluto's status. You didn't have to be a kid to feel this way. Craig Manister, a full-grown acquaintance, muttered to me at a cocktail party, "It's like knowing when you get out of bed that the floor is solid."

Over the years, I noticed a trend in the mail to my office from elementary schools. Slowly, legions of angry students move on, making room for new crops of students who may have never known the certainty of nine planets in the first place.

In a stack of letters sent in March 2005 from Mrs. Chemai Gray's middle school class in Marysville, Washington, the split of votes for Pluto's planet status had reached 50 percent in favor, 50 percent against, with students tending to offer arguments about size and tradition. A year later, other stacks of letters, including a second stack from the same teacher, showed that the students had become fluent in the Kuiper belt and the difference between icy and rocky objects in the solar system. They further gleaned a basic understanding of circular and elliptical orbits, and their correspondence was mostly devoid of emotion and sentimentality. By the end of 2006, letter tallies approached 90 percent against planet status and 10 percent in favor.

Meanwhile, others with opinions could not resist e-mailing me their commentary.

From the downright grumpy:

Dear Natural History Museum.
You are missing planet Pluto. Please make a model of it. This is what it looks like.
It is a planet.
Love
Will Galmot
Turn to the other page

Figure 5.1. Letter from Will Galmot.

Date: Feb 13, 2005, at 10:50 PM

I do not appreciate your attempting to demote Pluto to a non-planet.

I am 59 years old. I grew up on Tom Corbett, Space Cadet. Although I have seen many changes in my life, one thing I am sure of is that there are nine planets in the Solar System, and that the smallest and most distant is Pluto.

Leave it alone.

Dan E. Burns

To "the child shall lead them":

Date: November 18, 2004 7:09:13 PM EST

My name is John Glidden. I am six years old and my favorite planet is Pluto. I disagree with you that Pluto is a Kuiper Belt Object. I think Pluto is a real planet and I took a poll of 11 people. The question was, What do you think Pluto is?

A Planet

A Double Planet

A Kuiper Belt Object

A planet and A Kuiper Belt Object

I think it is a double planet and everyone else thought it is a regular real planet that is very cold.

I had a half day at school yesterday so my mom brought me to the Museum of Natural History and the Hayden Planetarium. I wanted to see you so I could tell you this in person.

John Glidden

To pleading with tongue-in-cheek arguments:

Date: Thu, 28 Jan 1999 07:45:58 -0500

C'mon Neil. Let's not go messing around with Pluto's planethood.

Grandfather the little guy and get on with it. (If Ceres starts to make a fuss, toss him an honorary something). Why, stripping Pluto of his planethood is like stripping George Washington of his citizenship because the US wasn't really a country when he was born.

In any case, the cost to make the change will be huge. There must be thousands of sets of twenty year old encyclopedias that will have to be replaced. The number of encyclopedia salesmen required to do it will surely tilt the social fabric of the world.

Steve Leece

To mild accusations of cultural insensitivity:

Date: December 6, 2004 9:06:50 PM EST

Would you say a small child or midget wasn't a person? Of course you wouldn't, although they are a different versions of the normal standard that is set as what a person would like, but they are still classified as people. By saying that Pluto is not a planet, is like saying a midget or a small child is not a person.

Brooke Abrams

And to blunt obstinacy:

Date: November 13, 2003 9:01:07 AM EST

Pluto is a planet because I say so. I don't think that something that I have been told all my life (namely, that Pluto is in fact a planet) should be doubted to the extreme that we must write a letter to some organization that does not in fact care what we think.

Lindsey Greene

Letters such as these provided months of entertaining reading, but little did I know, the assault had only just begun.

THE INTERNATIONAL ASTRONOMICAL UNION AND THE LITTLE PRINCE COME TO BLOWS OVE
NEWLY PROPOSED PLANETARY DESIGNATIONS.

6

Pluto's Judgment Day

After two years of committee deliberation, the International Astronomical Union (IAU) could not converge on what exactly a planet should be. And so the IAU formed an ad hoc Planet Definition Committee to see if they could succeed where others had failed. This fresh group of seven—five scientists, a journalist, and a science historian—met for two days before the Prague meeting to cogitate and then decided on what, in their judgment, would be the best solution for all concerned

Figure 6.1. *The seven-member Planet Definition Committee of the International Astronomical Union (IAU).* Top row, left to right: *André Brahic, University of Denis Diderot (Paris) planetary scientist and popularizer; Iwan Williams, Queen Mary University (London) planetary theorist; Junichi Watanabe, director of the Outreach Division of the National Astronomical Observatories of Japan; and Richard Binzel, MIT planetary scientist.* Bottom row, left to right: *Catherine Cesarsky, director general of the European Southern Observatories and president-elect of the IAU; Dava Sobel, best-selling science writer and journalist; and Owen Gingerich, Harvard astronomer, historian of science, and chair of the committee.*

parties, Pluto included. On August 16, 2006, they recommended to IAU membership that a planet be officially defined as an object that (1) is in orbit around a star, but not in orbit around another planet, and (2) is large enough for its own force of gravity to shape it into a sphere, but not so large that it would trigger fusion in its core becoming a star. This definition would have kept Pluto as a planet and added, on the spot, three more objects to the planet list: Ceres, Charon, and Eris, with many more surely to come.

In spite of the undeniable cosmic expertise represented among the committee members, they were nonetheless absent researchers who specialize in the discovery and analysis of Kuiper belt objects or in the discovery and analysis of exosolar planets, two frontiers of planetary science that bring daily insight to what kind of solar system we live in. Based on the comments and reactions already expressed by Kuiper belt codiscoverers David Jewitt and Jane Luu, for example, had either of them been on the committee, it surely would have led to yet another hung jury.

During the week that passed between the IAU proposal going public and the formal vote on the recommendations, the roundness criterion received substantial media attention. In an appearance on Comedy Central's *The Colbert Report*, I shared this information with faux ultraconservative host Stephen Colbert, who had been supportive of Pluto all along, but was deeply concerned that if being round was what made you a planet, then "that means anything could be a planet," and "if everything is a planet, then nothing is a planet." His concern, shared by many, was simply that the decision was "taking away the specialness of Earth's planetness." He then proceeded to trash-talk the other three planet candidates in the solar system, beginning with Pluto's very round moon Charon:

> *Hey Charon, you're orbit is so big, you only get Christmas once every 248 years, even then all you get is earmuffs because it's so cold!*

Next came Ceres, the largest and only round asteroid:

> *Hey Ceres, guess what? They call you a planet, but we both know you're just a big fat ass-teroid. Yeah. You're so ugly, God tried to hide you in an asteroid belt!*

Last was the yet-to-be-named icy Kuiper belt object 2003 UB313 that would later become Eris:

> *Hey 2003 UB313, if that is your real name, you're not a planet, you're just a lazy comet. Your mama's so ugly, she named you 2003 UB313.*

Back in the real world, the conference attendees hotly debated the roundness criterion for planethood, leading to two additional criteria: (1) that the round object not be in orbit around another, larger world—precluding Charon from being called Pluto's companion planet; and (2) that the round object has cleared its orbit of wayward debris—the death knell for Pluto, whose orbital regime remains rich with countless thousands of craggy chunks of icy Kuiper belt objects. This leaves the Sun with an eight-planet family instead of either twelve or nine. Coincidentally, on August 16, 2006, my friend and museum colleague Steven Soter (the fellow who first called my attention to the Pluto problem back in 1998) submitted for publication a research paper titled "What Is a Planet?" in which he quantifies what it would mean for an object to clean its orbit.[33] This criterion is subtle because without a quantitative account of a clean orbit, the requirement can be arbitrarily invoked. For example, as noted earlier, Earth continues to plow through hundreds of tons of wayward meteoroids per day in its annual journey around the Sun. So have we cleared our orbit? Clearly not. The objective is to assess the total mass of cleanable debris and compare it with the mass of the planet in question. If the debris does not amount to much, then you can claim to have cleaned or dominated your orbit. Otherwise, you're just one of the crowd.

For example, Earth far outweighs the sum of all matter it will ever collide with. Earth can plow through its daily dose of debris for a quadrillion (1,000,000,000,000,000) years and become a mere 2 percent heavier than when it started. A quadrillion years is 10,000 times longer than the current age of the universe. Meanwhile, the countless Kuiper belt comets outweigh Pluto by a factor of at least 15.

Steve Soter's paper provided perspective on what was, at the time, a

33. Steven Soter, "What Is a Planet?" *Astronomical Journal* 132 (2006): 2513. Also see "What Is a Planet?" *Scientific American* 296, no. 1 (January 2007): 20–27.

Figure 6.2. *The 26th (triennial) General Assembly of the International Astronomical Union, held in Prague. Of the 2,500 attending members, 424 remained for the last day of the conference (August 24, 2006) to vote overwhelmingly (90 percent in favor) on a revised definition of the word* planet *that excluded Pluto, formally "demoting" it to dwarf planet status.*

hastily added criterion to the original roundness definition of the IAU resolution. Soter and I had collaborated on this paper during its early stages, but by the end, he had done 95 percent of the work while I was (regrettably) distracted by administrative matters. So I withdrew as coauthor but was delighted to be recognized in the paper's acknowledgments.

Back at the IAU conference in Prague, anxious reporters waited outside the assembly hall, which was off-limits to the press. They lurked with the kind of silent anticipation one finds only during the election of a new pope by the college of cardinals, for which eager onlookers in St. Peter's Square search for the smoke to rise from the Vatican Palace chimney—black smoke, a failed ballot; white smoke, a new pope is elected. All week long my e-mail in-box logged more than a hundred inquiries a day on Pluto alone, from concerned citizens and from the press wanting me to comment. When the final vote was cast on August 24, 2006, a revised definition of a planet emerged—and a revised status for Pluto:

Pluto is officially demoted to the status of "dwarf planet."

More than 90 percent of the 424 voters voted for demotion. (See the full amended resolution 5A in Appendix F.) The same criteria that downgraded Pluto elevated Ceres to the class of dwarf planet from the ranks of asteroids. Eris, the freshly discovered spherical Kuiper belt object, joined Pluto among the ranks of dwarf planets as well.

By now, the Pluto e-mails to my in-box were arriving at about two hundred a day, with assorted subject headers that betrayed the sentiments contained within: "Now Look What You've Done!" "Congratulations on Winning the Pluto Is Not a Planet Debate," and "Honk if you think Pluto is still a planet." And a dozen major media outlets had called or e-mailed for my reaction to the decision. I happened to be at the beach that entire week, on vacation with my family, and could take no interviews. So this feeding frenzy would have to happen without me.

In the weeks that followed, there were one or two supporters amid the e-mail barrage:

Date: October 27, 2006 2:56:24 PM EDT

I was compelled to write . . . after hearing what I perceived as a breath of fresh analytic air, after spending so much time reading what amounted to philosophical flatulence.

Ian Stocks, Clemson University

The following comment was not sent directly to me but was posted to "Dome-L," an Internet chat group that serves planetarium professionals. The fellow reflects on the planetarium community's resistance to our museum's treatment of Pluto on the grounds that it was against IAU proclamation; but after the IAU officially demoted Pluto, the same community continued to object, this time ignoring the IAU. Behavior such as this betrays hidden biases that do not tend to be subject to rational argument.

Date: August 31, 2006 3:36:12 PM EDT

To: Dome-L

Pardon my insolence, but I'm mighty amused at some of the

responses to the IAU's decision. Particularly, I'm quite tickled that some of the most irate are the same people who decried Neil Tyson when he omitted Pluto from the solar system exhibit at AMNH.

Isn't it an odd twist that those who derided Tyson for flying in the face of Pluto's planetary status as granted by the IAU by omitting it (then) are the very same who are now assuring all who'll listen that they'll still be referring to Pluto as a planet in their planetarium programs!

Seems a bit hypocritical to me. . . .

Michael J. Narlock

Some people got a bit carried away in their anti-Pluto enthusiasm:

Date: August 27, 2006 2:48:26 PM EDT

F##k pluto, it was a sorry excuse for a planet anyhow, good riddance to bad solar trash, but now that the name is free, why not rename Uranus Pluto and get rid of all that grade school snickering.

Howard Brenner

Others took the occasion to lambaste the negotiating talents of well-meaning scientists:

Date: 07:40 AM 8/27/2006:

This whole issue is a marvelous example of why scientists/technocrats generally make poor politicians.

Dave Herald, Canberra, Australia

Meanwhile, surface mail continued. Angry third graders from the year 2000 were now in high school, with other (hormonal) priorities to distract them. But, as already noted, there's always a new crop of elementary schoolers to fill that void. In a stack of letters addressed to me from Mrs. Debbie Dalton's third-grade class in the Warren L. Miller Elementary School, in Mansfield, Pennsylvania, Emerson York expresses their sentiment best, complete with seven exclamation marks to end the letter, followed by an illustration of a teary-eyed Pluto (Figure 6.3).

One of my favorites of the angry-kid genre arrived from Madeline Trost, of Plantation, Florida, and was mailed on September 19, 2006. After addressing the envelope to me personally, she bluntly addresses her letter "Dear Scientest" (Figure 6.4), and she can't contain her flurry of assaults on my integrity, ending with an appeal to accommodate a shortcoming of her own. I received another angry letter from a kid, except this one was a little older. As a card-carrying member of the American Museum of Natural History, she also felt comfortable schooling me in Plutonian mythology (Figure 6.5).

For newspaper articles, journalists hardly ever get to title their own piece. That task usually goes to someone else in the back office. For Pluto and its demotion, the urge to compose an attention-getting headline that poked fun at the entire episode was irresistible—especially for the tabloids. Ones that rise above others include the *Atlanta Journal-Constitution*'s "Planned Planethood" on August 16, 2006. Meanwhile, with the fresh memory of miscounted ballots in Florida during the hotly contested presidential elections of 2000, the *St. Petersburg Times* carried the headline "Pluto's Hanging Chad" on August 22, 2006.

Those were the real headlines.

In a parody of *The New York Times*' page 1, The People's Cube (The PeoplesCube.com) ran a series of Pluto-inspired headlines (accompanied by illegible columns of text) that mirror prevailing political and cultural sentiments in America. Dated August 26, 2006, the page begins with:

Figure 6.3. *Letter from Emerson York.*

November 6, 06

Dear MR. Tyson,

I think pluto is a planet. Why do you think pluto is no longer a planet? I do not like your anser!!! Pluto is my faveret planet!!! You are going to have to take all of the books away and change them.

Pluto IS a planet!!!!!!

your friend
Emerson York

Figure 6.4. *Letter from Madeline Trost.*

Dear Scientest,

What do you call pluto if its not a Planet
anymore? If you make it a planet agian all the
Science books will be right. Do Poeple live
On Pluto? If there are Poeple who live
there they won't exist. Why can't
Pluto be a Planet? If its small
doesn't mean that it doent have to
be a Planet anymore. Some Poeple
like pluto. If it doen't exist then they
don't have a favorite Planet. Please write
back, but not in cursive because I can't read in
cursive.

Your friend,
Madeline Trost

DR. NEIL DEGRASSE TYSON
DIRECTOR, HAYDEN PLANETARIUM

9/7/06

DEAR DR. TYSON,

SEE WHAT YOUSE GUYS DONE DID TO US KIDS? (EVEN A 72-YEAR-OLD LIKE ME!)
PLUTO LIVES!!

Diane R Kline
Member, Museum of Natural History

P.S. IF PLUTO WERE IN EARTH'S ORBIT IT WOULD GROW A TAIL AND BE CALLED A COMET???? PLUTO HAS A MOON!! WHAT DO YOU FELLOWS PROPOSE WE DO WITH CHARON? CONSIGN IT TO HADES??? PROSERPINA NEEDS A NEW BALL TO PLAY WITH??

Figure 6.5. *Letter from Diane Kline.*

The New York Times: Pluto Crisis Edition

We quickly see residual anti-Islamic sentiment from September 11, 2001.

Muslim Protesters Burn Local Planetarium "Just in Case"

We then learn that Congress might have had something to do with the demotion.

Lack of Federal Funding Leads to Downsizing of Solar System

And we further learn the potential impact of Pluto's demotion elsewhere in the galaxy.

Pluto Decision Sends Shockwaves to Neighboring Solar Systems

The news of the day would not be complete without partisan politics

REPUBLICANS DENY AID TO PLUTO AMIDST GROWING CONCERNS FOR THE FUTURE OF TRANS-NEPTUNIAN OBJECTS

CAN PLUTO OUSTING HELP DEMOCRATS WIN ELECTIONS? AL GORE DEMANDS A RECOUNT OF ASTEROIDS

and Bush accusations.

NASA: BUSH KNEW ABOUT PLUTO'S INSUFFICIENT GRAVITY: ORDER TO "OUT" PLUTO MAY HAVE COME FROM KARL ROVE

America's strained relations with Venezuela did not go unnoticed either.

HUGO CHAVEZ PLEDGES TO SEND OIL TO PLUTO

You can never ignore any part of the Middle East.

IRAN PRESIDENT DEFENDS PLUTO, THREATENS TO RETALIATE AGAINST ISRAEL

HAMAS LEADERS TO APPEAL TO UN AS SOON AS THEY FIND OUT WHAT PLUTO IS

HEZBOLLAH CLAIMS ROCKETS CAN NOW REACH PLUTO

There's immigration-inspired politics, too.

MCCAIN TO GRANT PLANETARY STATUS TO ASTEROIDS IF ELECTED

And persistent claims of discrimination.

"BIG-PLANETISM" RAMPANT AT NATIONAL OBSERVATORIES: WHISTLEBLOWER UNCOVERS BIAS TOWARDS SMALLER, "FEMALE" PLANETS

Related, but smaller headlines follow.

REPUBLICANS SHRUG OFF GLASS CEILING FOR DWARFS, ASTEROIDS

POLL: MOST AMERICANS THINK
THAT BLACK HOLES ARE DISCRIMINATED AGAINST

PLUTO RULING ANGERS DWARVES, MIDGETS:
CLASS ACTION "DWARF TOSSING" LAWSUIT FILED

These headlines are not simply parodies of news but mirrors to the mores of modern America.

Newspapers also serve as the daily repository of public sentiment through their op-ed pages and their letters to the editor. In the *Houston Chronicle* of September 3, 2006, Randi Light wrote: "Pluto was voted out as a planet by a group of astronomers. But I've heard that Pluto will run as an independent now." The *Oregonian* from the same day printed a rhyme by Mike Malter, of Southwest Portland:

Dear Pluto,

The news is quite bad
Your recent demotion so sad

Earth scientists morphed you
They downsized and "dwarfed" you
Little buddy, I think you've been had.

Deeply concerned for the demise of America, Gene Lolnowski, of Ellicottt City, Maryland, wrote in the August 31, 2006, *USA Today*: "Our traditional values in this country are taking a big enough beating, and now the IAU wants to mess around with the traditional organization of the solar system. When is this going to stop? This Pluto decision must be reversed. Tradition must prevail."

Deeply concerned for Pluto's emotional stability, Marla Warren, from Bartonville, Illinois, wrote in the *New York Times* of August 28, 2006: "I can accept the rationale for stripping Pluto of its planet status. But was it necessary to stigmatize Pluto with a negative label about its appearance?

Calling any heavenly body 'dwarf' could very well damage its self-esteem. I propose a more positive classification; for example, assistant planet, apprentice planet or, perhaps, training planet."

In spite of widespread accusations to the contrary, I had no vested interest in the outcome of the IAU vote. As already noted, the solar system exhibits at the museum's $230 million Rose Center for Earth and Space in New York City did not organize objects by whether or not they were formally classified as planets. So the design and concept was largely immune to what was decided in Prague.

I remind the reader that the IAU does not normally vote on scientific concepts, heated or otherwise. Voting typically addresses noncontroversial things like nomenclature that clarifies or unifies our means of communicating with one another. Science is not a democracy. As is often cited (and attributed to Galileo), the stated authority of a thousand is not worth the humble reasoning of a single individual. Yet the IAU's vote to demote Pluto sure looked like an attempt at democracy. Immediately following the vote, many in the planetary science community protested. Some did so on grounds that the 424 voting astrophysicists could not possibly represent the 2,000+ astrophysicists who attended the conference or the 10,000+ world membership of the IAU. Others complained about the limited time made available for the community to mull over the draft resolutions.

Still others—actually, the effort was led by Mark Sykes (see Figure 4.8)—instantly circulated an online petition to allow the international community of scientists to protest the IAU vote if they chose to do so. Reprinted in its entirety below, the petition's text was a model of simplicity:[34]

34. Petition Protesting the IAU Planet Definition; http://www.ipetitions.com/petition/planetprotest.

Petition Protesting the IAU Planet Definition

> *We, as planetary scientists and astronomers, do not agree with the IAU's definition of a planet, nor will we use it. A better definition is needed.*

Open to signatures for the five days that followed the IAU vote, 304 scientists joined the list. The next day, August 31, 2006, the petitioners issued a press release with a swords-drawn opening line:

> *Sufficient signatures from planetary scientists and astronomers have been gathered to bring into serious question the definition for planet adopted by the IAU as fundamentally flawed, as was the process by which it was generated.*

After a long paragraph citing the impressive planetary pedigree of the signers, the press release called for a new, grassroots, inclusive effort to establish the definition of a planet. The process would culminate in a conference, "not to determine a winner, but to acknowledge a consensus." The release was signed by Arizona's Planetary Science Institute and Colorado's Southwest Research Institute.

Who knows what will ultimately become of this petition? More people voted against Pluto at the 2006 IAU Prague assembly than signed the petition itself. Of widespread concern to the petitioners was that 424 voting attendees amounted to a mere 4 percent of the world's astrophysicists, so how could the tally possibly represent the informed judgments of the entire community? On the surface, this argument sounds convincing, but most pollsters would give their eyeteeth for their sample to represent 4 percent of a complete population.

So the question should instead be, What are the chances that the vote would be substantially different if you polled all the world's astrophysicists? It turns out, if you do the math, that the vote's margin of uncertainty is less than 3 percent, which means that there is a 95 percent (2 sigma, in statistical parlance) chance that if the entire population were polled, the vote would fall within 3 percent of the tally obtained in Prague. The calculation assumes that the 424 scientists are a random sample. There is no reason to presume otherwise, except that people who favor Pluto's planethood typically exhibit more energy for their cause than Pluto demoters exhibit for theirs. So the 90 percent who voted for demotion may actually be lower than what one might expect for the entire population.

Here's another way to look at the problem: Suppose the people who signed the petition do not overlap at all with the 10 percent of the 424 who voted for Pluto's planet status in Prague. This is certainly not the case, but it offers an important, extreme view on what the numbers can tell us. Only 42 people voted for Pluto in Prague. Add that to the 304 who signed the petition, and we get about 350 professional Pluto-is-a-planet supporters worldwide. This figure is a mere 3.5 percent of the world's astrophysicists. Of course, not voting for something is not the same as voting against it. Most astrophysicists probably don't care enough about the problem to express an opinion at all. As suggested in Chapter 2, where I chronicle Pluto's disproportionate grip on the hearts and souls of the American public, the effect seems to be true for professionals as well. No more than 20 signers of Sykes's petition (about 6 percent), which circulated internationally, hailed from non-American institutions. Yet non-Americans comprise more than two-thirds of the IAU membership.[35]

This analysis notwithstanding, rather than pit petitions against votes, what should happen, and what Sykes calls for, is the search for consensus. And until one is obtained, nobody should be defining anything.

35. "IAU Geographical Distributions of Individual Members"; http://193.49.4.189/Geographical_Distribution.8.0.html.

SEE PLUTO
VICTIM OF DOWNSIZING
chas almon

7

Pluto the Dwarf Planet

THE WORLD COULD NOT STOP REACTING TO PLUTO'S new dwarf planet status. As though August 25, 2006, the day after the International Astronomical Union's vote, heralded a new zero point on the planetary calendar, giving us BD (Before Dwarf) for all dates prior and AD (After Dwarf) for all dates afterward.

After the IAU vote, Bill Nye immediately returned to my e-mail in-box with further critical observations of the way things were going:

The current International Astronomical Union (IAU) proposal to refer to Pluto as a "dwarf planet" will not be useful, because the word "planet" appears in a designation that is intended to explain that bodies like Pluto are not planets—a remarkable failure of a committee trying perhaps to please too many people.

Bill was not alone in his sentiment, although this widespread meaning was not the committee's intent. They added the word "dwarf" the way astrophysicists have used it for dwarf galaxy (which is still a galaxy) and for dwarf star (which is still a star). But to no avail. As far as anyone was concerned, the IAU killed planet Pluto.

During the demotion commotion, singer-songwriter Jonathan Coulton posted an ode to Pluto as sung by a loving Charon, titled "I'm Your Moon" (see complete lyrics in Appendix C).[36] The song opens with reference to Pluto's lack of a ring system. While not a criterion for planethood, Coulton was just warming up:

They invented a reason.
That's why it stings.
They don't think you matter
Because you don't have pretty rings.

He then poetically indicts the cavalier behavior of warring astronomers:

Let them shuffle the numbers.
Watch them come and go.
We're the ones who are out here,
Out past the edge of what they know.

36. Jonathan Coulton, "I'm Your Moon," 2006; http://www.jonathancoulton.com/songdetails/I'm%Your%Moon.

The refrain captures the important fact that among moons in the solar system, Charon and Pluto come closest to each other in size, allowing Charon to affectionately think of Pluto as its moon too:

I'm your moon.
You're my moon.
We go round and round.
From out here,
It's the rest of the world
That looks so small.

Amid the romance, Coulton offers a blunt reality check:

Sad excuse for a sunrise.
It's so cold out here.
Icy silence and dark skies
As we go round another year.

My favorite part of the song resembles what might transpire in a self-help therapy session:

Promise me you will always remember
Who you are.
Who you were.
Long before they said you were no more.

These are surely the most sensitive words ever shared between two inanimate cosmic objects.

In another Pluto-inspired song, going by the plain and simple title of "Pluto's Not a Planet Anymore" (for complete lyrics, see Appendix D), Jeff Mondak collaborated with Alex Stangl.[37] Mr. Mondak lives in

37. Jeff Mondak and Alex Stangl, "Pluto's Not a Planet Anymore"; http://jeffspoemsforkids.com.

Champaign, Illinois, where he is a children's poet and songwriter and a professor at the University of Illinois. Mr. Stangl lives in Peterborough, Ontario, where he is a singer, songwriter, musician, and music producer. This duo had collaborated on several songs before. They wrote and composed "Pluto's Not a Planet Anymore" on the suggestion of students at Barkstall Elementary School, in Champaign.

The song is upbeat, invokes catchy phrases, and revisits the line "Pluto's Not a Planet Anymore" enough times that you can just hear a classroom of kids belting it out in unified chorus. Here are my two favorite stanzas:

Uranus may be famous
But Mercury's feeling hot
For Pluto was a planet,
And somehow now it's not

Neptune's nervous, Saturn's sad,
And jumpin' Jupiter is hoppin' mad
Eight remain of nine we had
Pluto's not a planet anymore

The song ends with a clever, simple rhyme:

They met in Prague and voted
Now Pluto's been demoted
Oh, Pluto's not a planet anymore

Apart from songwriters inspired by personified dwarf planets, the next best sign that an obscure subject, or any subject at all, has entered the realm of pop culture is when that same subject becomes comedic fodder for humorists. A joke is funny only when everybody already knows the foundations of its content, allowing the writer to offer fresh comedic vistas without the burden of establishing context. Is there anything knee-

Figure 7.1. Political cartoonist Bob Englehart, of the Hartford Courant, *chose to exploit the "farthest planet" contest by making a larger political statement.*

slapping about the planet Mercury? Or Neptune? Or Alpha Centauri, the nearest star system to the Sun? Can't say I've ever heard a joke about any of them. But what humorist could possibly resist the parody of learned scientists carrying on like children as they argue about Pluto's status? And be they humorists or not, who could resist the playful personification of Pluto: simultaneously a planet, a nonplanet, a dog, an underdog, and an ice ball?

As we have already seen with media headlines, Pluto's demotion became a window on who and what we are as a culture, blending themes drawn from party politics, social protest, celebrity worship, economic indicators, academic dogma, education policy, social bigotry, and jingoism.

As if regional lawmakers had nothing better to do with their time, at least two state legislatures decided to take the Pluto problem into their own hands. New Mexico is longtime home of Pluto discoverer Clyde Tombaugh and, with its clear nighttime skies, location to world-class astronomical facilities, including the Apache Point Observatory, the Very Large Array, the Magdalena Ridge Observatory, and the National Solar Observatory (located by the way in the town of Sunspot, New Mexico). The state legislature felt that the IAU had unjustifiably dissed Pluto and, by association, their great state. So on March 8, 2007, their 48th Legislature, in a bill introduced by Representative Joni Marie Gutierrez, passed a Joint Memorial declaring Pluto a planet within state borders and making March 13, 2007, "Pluto Planet Day" statewide. (See Appendix G for the full text.)

The bill is not all grumpy complaints. Under several of the various paragraphs that begin with the ubiquitous "Whereas . . . ," one learns a bit of astronomy in the process:

> **WHEREAS**, *Pluto has been recognized as a planet for seventy-five years; and*
>
> **WHEREAS**, *Pluto's average orbit is three billion six hundred ninety-five million nine hundred fifty thousand miles from the sun, and its diameter is approximately one thousand four hundred twenty-one miles; and*
>
> **WHEREAS**, *Pluto has three moons known as Charon, Nix and Hydra; and*
>
> **WHEREAS**, *a spacecraft called* New Horizons *was launched in January 2006 to explore Pluto in the year 2015;*

What I don't know is this: if I shout "Pluto is *not* a Planet !" in a public theater in New Mexico, could I get arrested?

California was apparently way ahead of New Mexico in Pluto legislation. Its state legislature had a bill ready to go on August 24, 2006, practically minutes after the demotional vote was cast in Prague. While it did not ultimately pass, the bill was enthusiastically introduced by Assembly Members Keith Richman and Joseph Canciamilla. Bill HR36 (see Appendix H for the full text) calls the International Astronomical Union "mean spirited" and formally condemns the IAU's decision to strip Pluto of its planetary status for its "tremendous impact" on the people of California and the state's "long term fiscal health."

Tremendous impact on the people of California? It's all there nestled within the multiple appearances of "Whereas . . .":

> **WHEREAS**, *Downgrading Pluto's status will cause psychological harm to some Californians who question their place in the universe and worry about the instability of universal constants;*

Fiscal health of California? That's there, too, couched in terms of the California educational system:

> **WHEREAS**, *The deletion of Pluto as a planet renders millions of text books, museum displays, and children's refrigerator art projects obsolete, and represents a substantial unfunded mandate that must be paid by dwindling Proposition 98 education funds, thereby harming California's children and widening its budget deficits;*

How about shady politics?

> **WHEREAS**, *The downgrading of Pluto reduces the number of planets available for legislative leaders to hide redistricting legislation and other inconvenient political reform measures;*

And then there's the matter of Mickey's pet:

> **WHEREAS**, *Pluto, named after the Roman God of the underworld and affectionately sharing the name of California's most famous animated dog, has a special connection to California history and culture;*

Unlike their own state's legislature, the Disney Company of Burbank, California, accepted Pluto's demotion to dwarf status with grace and aplomb. In an official internally distributed memo titled "Despite Planetary Downgrade, Pluto Is Still Disney's 'Dog Star,'" apparently issued by the Seven Dwarfs (who have been dwarfs from the beginning), they console Pluto in his time of need:[38]

> *Although we think it's DOPEY that Pluto has been downgraded to a dwarf planet, which has made some people GRUMPY and others just SLEEPY, we are not BASHFUL in saying we would be HAPPY if Disney's Pluto would join us as an 8th dwarf. We think this is just what the DOC ordered and is nothing to SNEEZE at.*

The release continues:

> *As Mickey Mouse's faithful companion, Pluto made his debut in 1930—the same year that scientists discovered what they believed was a ninth planet. Said a whitegloved, yellow-shoed source close to Disney's top dog, "I think the whole thing is goofy. Pluto has never been interested in astronomy before, other than maybe an occasional howl at the moon."*

38. "Despite Planetary Downgrade, Pluto Is Still Disney's 'Dog Star,' " PR Newswire, August 24, 2006; http://www.prnewswire.com.

Remember that unlike canine Goofy, canine Pluto is a pet and so does not speak: hence the reference to Pluto howling at the Moon rather than offering an informed reaction to an inquiring press.

To Northeasterners, Californians have always looked (and behaved) a bit odd. In the days that followed the IAU vote to demote Pluto, Caltech media reported on a parade of a different kind through the streets of Pasadena:[39]

FUNERAL FOR A PLANET

Their heads hung low, accompanied by black-clad mourners and a jazz band, eight planets marched in a New Orleans–style funeral procession for Pluto in the 30th annual Pasadena Doo Dah Parade. They were joined by more than 1,500 parade participants, among which were the Marching Lumberjacks, guru Yogi Ramesh, Raelian devotees, the Zorthian nymph snake sisters, and the Men of Leisure and their Synchronized Napping Team, who stopped every now and then to recline.

The parade participants were mourning the open casket carrying Pluto:

Marching Lumberjack Karolyn Wyneken, who drove 700 miles from Humboldt County for the event, exclaimed, "Wow, that is awesome! That is so good, and necessary," upon seeing the open casket with its papier-mâché Pluto.

39. California Institute of Technology, "Funeral for a Planet"; http://pr.caltech.edu/periodicals/EandS/articles/LXIX4/funeral.html.

***Figure** 7.2. Varoujan Gorjian, who works on the Spitzer Space Telescope team at the Jet Propulsion Laboratories, marched in Pasadena's Pluto memorial parade as the red planet.*

Each planet in the precession was played by a different member of the Caltech community. The report continues, with a nepotistic account of Saturn and Earth:

> *Saturn, played by JPL postdoc Angelle Tanner and accompanied by her many rings, organized the march and voiced the sentiments of most of her fellow planets when she noted, "Most astronomers don't think Pluto should be a planet, but we all miss it." Some planets,*

however, felt strong-armed into participation—as trumpet-playing Earth (Samantha Lawler) noted, Saturn was "writing my recommendation letters."

Caltech Professor of Planetary Astronomy Mike Brown was also there, of course, accompanied by his daughter Lilah, who portrayed Eris, Brown's newly discovered queen of the Kuiper belt.

On August 17, 2006, Brian O'Neill, of the *Pittsburgh Post-Gazette*, invented a behind-the-scenes account of Pluto's demotion under the title "We See It as an Opportunity, Pluto." Here he imagines a conversation between Pluto and the astronomer-manager who breaks the news to him:[40]

"Hey, Pluto, thanks for coming in today. Have a seat."

"No, thanks. I'd rather hover."

"Well, Ploot—I think I can call you 'Ploot'—we're going to make some changes in the solar system, and you're going to be a big part of them."

"Great. Anything I can do for you guys in the white lab coats. I was just telling Neptune on my way past him a couple of hundred years back that we'd be nowhere without . . ."

"Yes, well, this is about you and Neptune and the others. A bunch of us in the International Astronomical Union got together and decided that, well, you're too special to be associated with the likes of Mercury and Mars."

40. Brian O'Neill, "We See It as an Opportunity, Pluto," *Pittsburgh Post-Gazette*, August 17, 2006; http://www.post-gazette.com/pg/06229/714139–155.stm.

You can imagine where this is going as the astronomer eases Pluto into the idea that he is different from the rest of his coworkers. The piece ends with an almost cliché reference to office politics:

> *"Look, Ploot, we recognize you're upset, but this is really just a lateral move, not a demotion. You're still a very important part of our solar system, and we're looking at other objects about your size that we may make part of your team."*

Some humorists felt compelled to parody cultural icons using the Pluto demotion story as a template. On MLB.com, the official Web site for Major League Baseball, Mark Newman, the enterprise editor for MLB.com, reported on the day of Pluto's demotion under the header "Pluto Sent Down to the Minors: Former planet hurt by lack of size, disgruntled fan base."[41] In a lengthy article that surely contains more science than has ever appeared on the MLB Web pages, Newman included a paragraph on planetary batting order, remembering that nine players as well as nine planets are what's supposed to constitute a team:

> *[Pluto] could never be Mercury, leading off and constantly hot. Venus was all about love and self-sacrifice, a natural 2 spot in the order. Earth, the prototypical No. 3 hitter, the ultimate fantasy pick, the people's choice. Mars, the oft-feared big red machine. Jupiter always had the sweet spot in the lineup. Having Saturn in the order always meant a ring. Uranus, always the team prankster and playing jokes to keep it fun. Year after year, Pluto tried to leap past Neptune at the end of the order. Because of its eccentric orbit, Pluto actually was able to reach closer to the sun than Neptune during a portion of its orbit.*

41. Mark Newman, "Pluto Sent Down to the Minors"; http://MLB.com/news.

But again and again, Neptune, the savvy veteran (discovered in 1846), would deny the kid. Pluto never really had a legitimate chance. The youngster with the cold streak also suffered from poor marketing.

Boston Globe sports columnist Dan Shaughnessy could not resist comparisons with Red Sox slugger Manuel "Manny" Ramirez. In August 27, 2006 Shaughnessy wrote:[42]

More news yesterday from the International Astronomical Union general assembly in Prague. In the wake of their controversial decision to demote Pluto, the astronomers have agreed to officially recognize Planet Manny as the new ninth celestial body in our solar system. Makes sense. Planet Manny operates in his own orbit and hits baseballs into outer space. He's certainly no dwarf planet like Pluto.

Continuing in the sports motif, the performance of the New York Knickerbocker basketball team (the "Knicks") had been so disappointing in September of 2006 that political humorist Andy Borowitz, of the online *Borowitz Report*, found reference to Pluto irresistible under the title "Scientists Say Knicks Are No Longer a Basketball Team: Prague Conference Demotes New York Team to Dwarf Status."[43] The short article makes good use of academic innuendo as a tool to convey the frustrations felt by all fans of the team:

Just weeks after a conference of scientists determined that Pluto was not a planet after all, the same scientists reconvened in Prague today

42. Dan Shaughnessy, "This Is One Star That Is in A Wobbly Orbit," *Boston Globe*, August 27, 2006.

43. Andy Borowitz, "Scientists Say Knicks Are No Longer a Basketball Team"; http://www.borowitzreport.com/archive_rpt.asp?rec=6582.

to pronounce that the New York Knicks were not a basketball team. Sports fans have suspected over the last few seasons that the original decision to characterize the Knicks as an actual NBA team may have been in error, but today's announcement by the scientists seemed to remove all remaining shreds of doubt.

From here onward, you could substitute Pluto for Knicks, and basketball team for planet, and get a sense of the actual scientific debate as it unfolded:

"While the New York Knicks possess some qualities that are consistent with a basketball team, we have come to the conclusion that they are something else entirely," said Dr. Hiroshi Kyosuke of the University of Tokyo." It would be more accurate to call the Knicks a dwarf team. "Dr. Kyosuke said it was "understandable" that scientists had assumed that the Knicks were a basketball team for so many years, because they exhibited behavior similar to such teams, such as moving around a basketball court in a seemingly organized manner and hurling an orange spherical object.

And here Borowitz can't be more blunt:

"However, they failed to exhibit two properties common to all basketball teams," Dr. Kyosuke said. "Scoring points and winning games." In New York, Knicks coach Isiah Thomas welcomed the reassessment of the Knicks, saying that being designated a dwarf team represented a unique opportunity for the franchise: "If this means that now we can play against actual dwarves, maybe we'll start winning."

Not limited to sports references, a month later, Andy Borowitz used the Pluto story to take a swipe at Washington, D.C., under the title "Scientists

Demote Bush Presidency to Dwarf Status: White House Joins Pluto in New Classification."[44] Taking his cue from the November 2006 elections results, in which the Republican White House lost control of Congress, Borowitz observed:

> *In the aftermath of the midterm elections . . . scientists called an emergency meeting in Oslo to determine if the Bush administration in fact still qualified as a presidency. . . . But with the president's approval rating in a free fall, it became clear even before the scientists convened that some sort of reclassification along the lines of the Pluto demotion was in order. . . . Dwarf status means that Mr. Bush is "less than a president, but more than a mayor."*

There's nothing quite like the free New York–based weekly *The Onion*. Billed as "America's Finest News Source," the newspaper's parodies are sharp, clever, hilarious, and written with such deadpan journalistic prose that half the time you find yourself double-checking to make sure that you had not accidentally picked up the *New York Times* or the *Washington Post*. In an article posted December 18, 2006, NASA was given the task of letting Pluto know of the IAU decision:[45]

> BEARER OF BAD NEWS
>
> *The* Consoler *probe braces to break the news to Pluto.*
>
> *"It's tough, but we thought giving it to Pluto straight was the right thing to do,"* NASA *Chief Engineer James Wood said. "After all,*

44. Andy Borowitz, "Scientists Demote Bush Presidency to Dwarf Status"; http://www.borowitzreport.com/archive_rpt.asp?rec=6632&srch=.

45. "NASA Launches Probe to Inform Pluto of Demotion" *The Onion*, no. 42.51, December 18, 2006.

> *it put in 76 years as our ninth planet—it just didn't seem fair to break the news with an impersonal radio transmission beamed from Earth."*
>
> *"Pluto is more than 3.5 billion miles from the sun,"* Wood said. *"Launching that probe felt like the best way to avoid alienating it any further."*

Appealing to modern-day issues regarding personal feelings and self-esteem, the article continues:

> Wood *said* Consoler *will "take pains" to explain to Pluto that the reasons for the demotion "had nothing to do with anything it did personally."*
>
> *Scientists at* NASA *have taken precautions that* word *of the demotion will not reach Pluto before* Consoler *does. The* New Horizons *probe, which will pass by Pluto in July 2015, has been instructed to maintain radio silence. It is, however, programmed to congratulate nearby Eris and Ceres for their promotion from asteroids to dwarf planets.*
>
> *"The* Consoler *probe will reach Pluto on a Friday, if our calculations are correct,"* Wood said. *"It's always better to do this kind of thing right before the weekend."*

Undaunted by the IAU vote, Maryn Smith, a 10-year-old fourth grader at Riverview Elementary School, in Great Falls, Montana, replied to a contest run by the National Geographic Society.[46] The task? To construct an 11-planet mnemonic, restoring Pluto to its rightful place in the pantheon of planets and boldly adding a word for the lone spherical asteroid Ceres between Mars and Jupiter and a word for Eris at the end. She won

46. News account reported by *Machinist*; http://machinist.salon.com/blog/2008/02/27/11_planet/index.html.

with the sentence "My Very Exciting Magic Carpet Just Sailed Under Nine Palace Elephants," citing the influence of Disney's *Aladdin*, and just in time for a book to be published by National Geographic titled *11 Planets: A New View of the Solar System*.[47] According to the Associated Press, singer-songwriter Lisa Loeb plans a song inspired by it, titled, of course, "My Very Exciting Magic Carpet."

Defiance at its finest.

While astrophysicists were downgrading the cosmic object we call Pluto, the American Dialect Society, which is more than a century old, was upgrading the status of the word Pluto to a verb, making it their 17th annual "Word of the Year" for 2006:[48]

> **to pluto / to be plutoed:** to demote or devalue someone or something, as happened to the former planet Pluto when the General Assembly of the International Astronomical Union decided Pluto no longer met its definition of a planet.

The society counts linguists, grammarians, and assorted scholars among its members, who vote for fun and not as part of an official edict. Their goal is to analyze the language, assess trends in usage, and then induct fresh and emergent words into the English language.

Dictionaries are sure to adopt the new word, given the many occasions in life that one could use the term. The word "plutoed" also enjoys rhyme and resonance with the similarly defined word "torpedoed."

47. David Aguilar, *11 Planets: A New View of the Solar System* (Washington, DC: National Geographic Children's Books, 2008).
48. American Dialect Society, "Plutoed"; http://www.americandialect.org/Word-of-the-Year_2006.pdf.

Not missing a beat, NBC's *Tonight Show* host Jay Leno reacted to the new word in his opening monologue on the night of January 19, 2007:

> *"I'm glad they chose Plutoed, instead of Uranused."*

The only way that joke works is to pronounce Uranus scatalogically as "your-anus," which of course Jay did.

Meanwhile, those people in society who would credit or blame the cosmos, and not themselves, for their financial affairs and love life were split on what impact an official statement to demote Pluto would have on their horoscope casting. The day after the IAU vote, a story in the *Wall Street Journal* by Jane Spencer appeared, under the title "Pluto's Demotion Divides Astrologers." The widely reprinted article cites the American Federation of Astrologers and the Astrological Association of Great Britain as standing firmly by Pluto, asserting that the icy orb is a full-blown planet, maintaining a powerful pull on our psyche, despite the IAU vote to the contrary. Then comes my favorite line:

> *"Whether he's a planet, an asteroid, or a radioactive matzo ball, Pluto has proven himself worthy of a permanent place in all horoscopes," says Shelley Ackerman, columnist for the spirituality Web site Beliefnet.com.*

The article goes on to quote Ms. Ackerman criticizing the IAU for not including astrologers in its decision. It further quotes Eric Francis, of Planetwaves.net, which represents a subgroup of these medieval prognosticators known as minor-planet astrologers: "This is a moment that I've been waiting for for a long time," Francis remarks as he welcomes Ceres, Eris, and Charon to the ranks of dwarf planets, granting horoscope charts extra ways for believers to cede control of their lives to the universe.

The article ends with *Vanity Fair* astrologer Michael Lutin saying that he will consider the newcomers, but remains skeptical of their influence on our daily affairs due to their location at the outer reaches of the solar system: "UB313 is never going to tell you whether Wednesday is good for romance." Actually, neither will anything else in the sky, unless it's an asteroid headed toward Earth, scheduled to hit on Wednesday.

Please tell your children to stay in school.

"SCIENCE CLASS JUST GOT A WHOLE LOT HARDER, KIDS..."

8

Pluto in the Elementary School Classroom

A Personal Recommendation for Educators

Yes, it really is official. Pluto is *not* a red-blooded planet, as voted in August 2006 by the general assembly of the International Astronomical Union. Pluto is now a "dwarf" planet.

How rude.

The vote overturned the Planet Definition Resolution proposed by the Planet Definition Committee, which had stated simply that round objects in orbit around the Sun are planets. Pluto is a round object. Therefore, Pluto is a planet. This first attempt to define "planet" would have given everyone the right to utter Pluto and Jupiter in the same breath even though Jupiter is 260,000 times larger than Pluto. Plutophiles had about a week to rejoice before learning the sad news that Pluto fails a new criterion—that a legitimate planet must also dominate the mass of its orbital zone. Poor Pluto is crowded by thousands of other icy bodies in the outer solar system.

Embarrassing as it was to us all, the term *planet* had not been formally defined since the time of ancient Greece.

In 1543, Nicolaus Copernicus published his thesis—his newfangled, Sun-centered (heliocentric) universe, which confounded the wanderer classification scheme. Instead of being stationary and in the middle of things, Earth moved around the Sun just like everybody else. From that moment on, the term *planet* had no official meaning at all. Astronomers just silently agreed that whatever orbits the Sun is a planet. And whatever orbits a planet is a moon.

Not a problem if cosmic discovery freezes in time. But shortly thereafter, we learned that comets orbit the Sun too, and are not, as long believed, local atmospheric phenomena. Are they planets too? No, we already had a name for them: comets. They're the icy objects on elongated orbits that throw a long tail of evaporated gases as they near the Sun.

How about the craggy chunks of rock and metal that orbit the Sun in a region between Mars and Jupiter? Hundreds of thousands roam there. Are they each a planet too? While first called planets—beginning in 1801 with the discovery of Ceres—it became rapidly clear as dozens more were discovered that this new community of objects deserved its own classification. They came to be called asteroids.

Meanwhile, Mercury Venus, Earth, and Mars form a family of their own, being relatively small and rocky, while Jupiter, Saturn, Uranus, and Neptune are large and gaseous, have many moons, and bear rings.

And what's going on beyond Neptune? Beginning in 1992, icy bodies were discovered that look and behave a lot like Pluto. Yet another swath of populated real estate was discovered, akin to the discovery of the asteroid belt two centuries before. Known as the Kuiper belt, in honor of the Dutch-born American astronomer Gerard Kuiper, who championed its existence, this region of the solar system contains Pluto, one of its largest members. But Pluto had been called a planet since its discovery in 1930. Should all Kuiper belt objects be called planets?

Without a formal definition for the word *planet*, these questions created years of debate among people for whom counting planets matters.

If my overstuffed in-box is any indication, planetary enumeration remains a major pastime of the elementary schools and a deep concern of the print and broadcast media. Counting planets is what allows you to invent clever mnemonics to remember them in sequence from the Sun, such as "My Very Educated Mother Just Served Us Nine Pizzas." Or its possible successor: "My Very Educated Mother Just Served Us Nachos." Here's one that may grow on us all: "My Very Educated Mother Just Said Uh-oh No Pluto."

Where do you go from there? Because of exercises such as this, elementary school curricula have unwittingly stunted an entire generation of children by teaching them that a memorized sequence of planet names is the path to understand the solar system. The word *planet* itself continues to garner profound significance in our hearts and minds. This was surely justifiable before telescopes let us observe planet atmospheres; before space probes landed on planet surfaces; before we learned that icy moons make fertile targets for astrobiologists; before we understood the history of asteroid and comet collisions. But today, the rote exercise of planet counting rings hollow and impedes the inquiry of a vastly richer landscape of science drawn from all that populates our cosmic environment.

Suppose other properties matter to you instead. Suppose you care about ring systems, or size, or mass, or composition, or weather, or state of matter, or proximity to the Sun, or formation history, or whether the cosmic object can sustain liquid water or liquid anything. These criteria represent demographic slices that reveal much more about an object's identity than whether or not its self-gravity made it round or whether or not it's the only one of its kind in the neighborhood.

Why not think of the solar system as families of objects with like properties, and the cut through these properties is yours to take. Interested in cyclones? You get to talk about the thick atmospheres of Earth and Jupiter together. How about auroras? That conversation would include Earth, Jupiter, and Saturn, since all three have magnetic fields that guide charged solar particles to their poles, rendering their atmospheres aglow. How about volcanoes? That would include Earth, Mars, Jupiter's moon Io, and Saturn's moon Titan, whose eruptions are likely driven by ice and not by lava. How about wayward orbits? That would have to include comets and near-Earth asteroids, themselves putting life on Earth at risk. The list of ways to envision and organize the solar system is long and likely endless.

Imagine a solar system curriculum that begins with the concept of density. A big idea for third graders, but not beyond their grasp. Rocks and metals have high density. Balloons and beach balls have low density. Divide the inner and outer planets in this way, as cosmic examples of high and low density. Have fun with Saturn, whose density, like that of a cork, is less than that of water; unlike any other object in the solar system, Saturn's material would float in a bathtub.

At no time are you counting things. At no time are you worried about the definition of a taxonomic category. At no time are you left in search of a mnemonic on the premise that to understand the solar system you must memorize the proper names for things.

Eventually, you might be curious about the joint criteria of roundness and isolation. It's undiscriminating enough to combine into the same category tiny, rocky, iron-rich Mercury and large, massive, gaseous Jupiter—at which point you remember that way back in August of the year 2006, the International Astronomical Union created a name for that class of object. You search the organization's archives, find the word *planet*, and then quickly move on to the rest of what piques your interest in the solar system.

9

Plutologue

At a July 2008 meeting in Oslo, Norway, the Executive Committee of the IAU approved a proposal[49] from the IAU Committee on Small Body Nomenclature for the name *plutoid* to classify all dwarf planets that one might find orbiting beyond Neptune. As of this printing, two dwarf planets orbit there—Pluto and Eris—with many more surely to come. This obviously makes

49. IAU press release IAU0804, July 11 2008; http://www.iau.org/public_press/news/release/iau0804/.

Pluto a plutoid. By IAU rules, Ceres, another dwarf planet in the solar system, does not receive this honor because it orbits between Mars and Jupiter in the asteroid belt. Odd that Pluto's moon Charon is excluded, as are any yet-to-be-discovered round moons of dwarf planets beyond Neptune. An arbitrary, but enforceable, rule.

Alan Stern, now of the University Space Research Association, does not like the plutoid class for many reasons astrophysical, but when interviewed, he typically leads with "It sounds like 'hemorrhoid.' "

He and I finally agree on something.

By August 2008, Pluto's pitbull, Mark Sykes, was at it again. He and others organized a conference for planetary experts to discuss what to do about classification schemes for the solar system, with little to no regard for IAU proclamations. Held at the Applied Physics Laboratory of Johns Hopkins University in Maryland, the conference included a heavily touted public forum titled "The Great Planet Debate."[50] Given my history with the subject, I felt somewhat accountable for things, so I agreed to debate Sykes himself, in a reprise of our impromptu conversation in my office six years earlier. But this time we had a moderator—NPR's *Science Friday* host, Ira Flatow. On entering the auditorium, Mark and I were showered with "Let's Get Ready To Rumble" entry music, commonly played as professional wrestlers enter the ring.

I am pleased to report that Mark was much more polite and cordial than I had ever remembered him to be. And at times, I was the one who was manic. But in the end, we did not converge on the definition of a planet. We both agreed, however, that the IAU had body-slammed Pluto on this one. And that a more enlightened solution to the problem awaited us all.

50. "Great Planet Debate: Science as Process," Mark V. Sykes and Neil deGrasse Tyson, moderated by Ira Flatow, John Hopkins University Applied Physics Laboratory, August 14, 2008; http://gpd.jhuapl.edu/

In an attempt to achieve closure, I took a pilgrimage to Orlando's Disney World. I felt duty-bound to alert Pluto (the dog) of my role in his demotion. After initial dismay, or at least what looked like dismay in a creature that cannot frown, Pluto and I became fast pals, and he has accepted his uncertain fate with grace and dignity.

Meanwhile, nobody is quite sure how Pluto (the ex-planet) feels about all this except, perhaps, cartoonists. Nearly 4 billion miles away, Pluto, a cosmic object by any name, offers the last word:

Dear Dr. neiL Degrasse
tysoh I Know
how you feel.
We feel the
same about plato
not being
a planet.
I,m to But
we Just
have to get
over it - thats
Science. We
to learn
about Science.
Science will
make you Smart!
Love! Siddiq age 8

PLUTO

Letter from Siddiq Canty, Mrs. Koch's second-grade class, Roland Lewis Elementary School, Tampa, Florida (spring 2008).

Appendix A

Pluto Data (2008)[51]

Discoverer	Clyde W. Tombaugh
Date of discovery	February 18, 1930
Mass (kg)	1.27×10^{22}
Mass (Earth = 1)	2.125×10^{3}
Equatorial radius (km)	1,137
Equatorial radius (Earth = 1)	0.1783
Average density (g/cm^3)	2.05
Average distance from the Sun (km)	5,913,520,000
Average distance from the Sun (Earth = 1)	39.5294
Rotational period (days)	–6.3872 [rotates backwards]
Orbital period (years)	248.54
Average orbital velocity (km/s)	4.74
Orbital eccentricity	0.2482
Tilt of axis (degrees)	122.52
Orbital inclination (degrees)	17.148
Equatorial surface gravity (m/s^2)	0.4
Equatorial escape velocity (km/s)	1.22
Visual geometric albedo	0.3
Visual magnitude (Vô)	15.12
Atmospheric composition	Methane, nitrogen

51. US Naval Observatory; http://aa.usno.navy.mil/data/; and Kenneth R. Lang, *Astrophysical Formulae*, vols. 1 and 2 (New York: Springer-Verlag, 1999).

Appendix B

"Planet X"

Complete lyrics © 1997, 2006 by Christine Lavin Music (ASCAP)
Administered by Bug Music
http://www.christinelavin.com
Reprinted with permission

In Arizona at the turn of the century,
astromathematician Percival Lowell
was searching for what he called "Planet X"
'cause he knew deep down in his soul
that an unseen gravitational presence
meant a new planet spinning in the air
joining the other eight already known
circling our sun up there.

But Percival Lowell died in 1916
his theory still only a theory
'til 1930, when Clyde Tombaugh
in a scientific query
discovered "Planet X"
3.7 billion miles from our sun
a smallish ball of frozen rock,
methane and nitrogen.

It joined Mercury, Venus,
Earth, Mars, Jupiter, Saturn, Uranus, and Neptune
our solar system's newest neighbor
two-thirds the size of our moon
a tiny, barely visible speck
Cold! Minus 440 below.
Not exactly Paradise,
they named the planet Pluto.

That same year, 1930, Walt Disney
debuted his own Pluto as well
but a cartoon dog with the very same name as the CEO of Hell
was not your normal Disney style
most figured he was riding the coattails
of Pluto-mania sweeping the land
(not unlike our modern love for dolphins and whales).

For the next five decades mysterious Pluto
captivated our minds
as late as 1978 its own moon Charon
was seen for the very first time
but now telescopes and satellites
and computer calculations
say that Pluto may not be a planet at all,
creating great consternation.

(Some scientists say)
That Pluto is a "trans-Neptunian interloper"
swept away by an unknown force
or a remnant of a wayward comet
somehow sucked off course
others say that Pluto is an asteroid

in the sun's gravitational pull
but if you ask Clyde Tombaugh
he'll tell you "That's all 'bull'."

"I get hundreds of letters from kids every year," he says,
"It's Pluto the planet they love.
It's not Pluto the comet,
It's not Pluto the asteroid
they wonder about above."
And at the International Astronomical Union Working Group
For Planetary System Nomenclature
They too say that Pluto is a planet
reinforcing Clyde Tombaugh's view of nature.

Norwegian Kaare Aksnes,
professor at the Theoretical Astrophysics Institute
He too says that Pluto is a planet
and a significant one, to boot
but at the University of Colorado
astronomer Larry Esposito
says "If Pluto were discovered today,
it would not be a planet. End of discussion. Finito."

He says that it was not spun off from solar matter
like the other eight planets we know.
By every scientific measurement we have
is Pluto a planet? No!
and now 20 astronomy textbooks
refer to Pluto as less than a planet
I guess if Pluto showed up at a planet convention
the bouncer at the door might have to ban it.

St. Christopher is looking down on all this
and he says, "Pluto, I can relate.
When I was demoted from sainthood
I gotta tell you little buddy,
it didn't feel real great"
and Scorpios look up in dismay
because Pluto rules their sign.
Is now reading their daily Horoscope
just a futile waste of time?

It takes 247 earth years
for Pluto to circle our sun.
It's tiny and it's cold
but of all heavenly bodies
it's Clyde Tombaugh's favorite one.
He's 90 now and works every day
in Las Cruces, New Mexico
determined to maintain the planetary status
of his beloved Pluto.

But how are we going to deal with it
if science comes up with the proof
that Pluto was never a planet.
How do we handle this truth?
As the Ph.D's all disagree
we don't know yet who's wrong or who's right
but wherever you are, whatever you are,
Pluto, we know you're out there tonight.

And in the year 2003
you're going to see
the NASA Pluto Express

fly by and take pictures
of your way cool surface
to send to this web page address:
h t t p colon slash slash d o s x x dot colorado dot edu slash
plutohome dot h t m l
You've got your own web page!
For a little guy,
you've made quite a splash!

Yes, at the turn of the 20th century
astromathematician Percival Lowell
in his quest for "Planet X"
started this ball to roll,
but at the end of the 20th Century
we think he may have been a little off base
so we look at the sky
and wonder what new surprises
await us in outer space.

Appendix C

"I'm Your Moon"

Complete lyrics © 2006 by Jonathan Coulton
http://www.jonathancoulton.com
Reprinted with permission

They invented a reason.
That's why it stings.
They don't think you matter
Because you don't have pretty rings.

I keep telling you I don't care
I keep saying there's one thing they can't change

I'm your moon.
You're my moon.
We go round and round.
From out here,
It's the rest of the world
That looks so small
Promise me you will always remember
Who you are.

Let them shuffle the numbers.
Watch them come and go.

We're the ones who are out here,
Out past the edge of what they know.

We can only be who we are
Doesn't matter if they don't understand

I'm your moon.
You're my moon.
We go round and round.
From out here,
It's the rest of the world
That looks so small.
Promise me you will always remember
Who you are.
Who you were
Long before they said you were no more.

Sad excuse fore a sunrise.
It's so cold out here.
Icy silence and dark skies
As we go round another year.

Let them think what they like, we're fine.
I will always be right here next to you

I'm your moon.
You're my moon.
We go round and round.
From out here, it's rest of the world
That looks so small
Promise me you will always remember
Who you are.

Appendix D

"Pluto's Not a Planet Anymore"

Complete lyrics © 2006 by Jeff Mondak and Alex Stangl
http://www.jeffspoemsforkids.com
Reprinted with permission

Since 1930, quite a run
It was always the smallest one,
And oh so distant from the sun
But Pluto's not a planet anymore

Astronomers who had a look
Said "go re-write your science book"
They gave it the celestial hook
Now Pluto's not a planet anymore

Listen James and Janet
Some experts said to can it
Now Pluto's not a planet
No, Pluto's not a planet
Anymore

Uranus may be famous
But Mercury's feeling hot
For Pluto was a planet,
And somehow now it's not

Neptune's nervous, Saturn's sad,
And jumpin' Jupiter is hoppin' mad
Eight remain of nine we had
Pluto's not a planet anymore

They held the meeting here on Earth
Mars and Venus proved their worth
But puny Pluto lacked the girth
So Pluto's not a planet anymore

Listen James and Janet
Some experts said to can it
Now Pluto's not a planet
No, Pluto's not a planet
Anymore

They met in Prague and voted
Now Pluto's been demoted
Oh, Pluto's not a planet anymore

Appendix E

Official Media Response from the Author Regarding the Rose Center's Exhibit Treatment of Pluto

Submitted to CCNet, UK-based scholarly Internet chat group, February 2, 2001

Regarding our exhibits in New York City's new Rose Center for Earth and Space, here at the American Museum of Natural History, I am surprised and impressed by the amount of recent media attention triggered by our decision to treat Pluto differently from the other planets in the solar system.

I am surprised because our exhibit has been in place since opening day, 19 February 2000, and our treatment didn't seem to be newsworthy at the time. I am impressed that people feel so strongly about Pluto that much time and attention had been devoted to it in print and on the air.

The *New York Times*' front page article, which ignited the recent firestorm, donned a title that was somewhat a-field of what we actually did, and which I would like to clarify. The title read "Pluto not a Planet? Only in New York," which implied that we kicked Pluto out of the solar system and that we are alone in this action and that, perhaps more humorously, Pluto wasn't big enough to make it in NYC.

I have written previously on the subject in an essay titled "Pluto's Honor" (*Natural History* Magazine, February 1999) where I review how the classification of "planet" in our solar system has changed many times, most notably with the 1801 discovery of the first of many new planets in orbit between Mars and Jupiter. These new planets, of course, later became known as asteroids. In the essay, arguing in part by analogy with the asteroid belt, I argued strongly that Pluto, being half ice by volume, should assume its rightful status as the King of the Kuiper belt of comets. Apart from my views expressed there, I have a different sort of responsibility to the public as director of the Hayden Planetarium and as project scientist of the Rose Center for Earth & Space.

That responsibility is as an educator for a facility that has received an average of 1,000 people per hour over the past eleven months.

For the exhibit on planets in our "Hall of the Universe," rather than use the word *planet* as a classifier, we all but abandon the ill-defined concept and simply group together families of like-objects. In other words, instead of counting planets or declaring what is a planet and what is not, we organize the objects of the solar system into five broad families: the terrestrial planets, the asteroid belt, the Jovian planets, the Kuiper belt, and the Oort cloud. With this approach, numbers do not matter and memorized facts about planets do not matter. What matters is an understanding of the structure and layout of the solar system. On other exhibit panels, in an exercise in comparative planetology, we highlight rings, storms, the greenhouse effect, surface features, and orbits with discussions that draw from all members of the solar system where interesting and relevant.

Our intro-exhibit panel meets the visitor's expectations head-on:

WHAT IS A PLANET?

In our solar system, planets are the major bodies orbiting the Sun. Because we cannot yet observe other planetary systems in similar detail, a universal definition of a planet has not emerged. In general, planets are massive enough for their gravity to make them spherical, but small enough to avoid nuclear fusion in their cores.

A second panel describes and depicts the layout of the solar system:

Our Planetary System

Five classes of objects orbit our Sun. The inner terrestrial planets are separated from the outer gas giant planets by the asteroid belt. Beyond the outer planets is the Kuiper Belt of comets, a disk of small icy worlds including Pluto. Much more distant, reaching a thousand times farther than Pluto, lies the Oort Cloud of comets.

Our goal was to get teachers, students, and the average visitor to leave our facility thinking about the solar system as a landscape of families rather than as an exercise in mnemonic recitation of planet sequences. We view this posture as the scientific and pedagogical high-road.

That being said, I have benefited from some reasoned feedback on what we have done. As many are already aware, we use our giant 87-foot sphere (housing the Hayden Planetarium Space Theater in the upper half and our creation of the first three minutes of the Big Bang in the lower half) as an exhibit unto itself. We invoke it to compare the relative sizes of things in the universe for a walk-around "powers of ten" journey that descends from the observable universe all the way to atomic nuclei. About midway in the journey you come upon the size scale where the sphere represents the Sun. There, hanging from the ceiling, are the Jovian planets (the most highly photographed spot in the facility) while a set of four small orbs sit on view, attached to the railing. These are the terrestrial planets. No other members of the solar system are represented here. This entire exhibit is about size, and not much else. But the absence of Pluto (even though the exhibit clearly states that it's the Jovian and Terrestrial planets that are represented) has led about ten percent of our visitors to wonder where it is.

In the interest of sound pedagogy we have decided to explore two paths: 1) Possibly add a sign at the right spot on the size scales exhibit that simply asks "Where's Pluto?" and gives some attention to why it was not included among the models. And 2) We are further considering a more in-depth treatment of the life and times of Pluto to add to our kiosks, which contain our computer-searchable data base of current astrophysics news that we display in a timely fashion on a video "bulletin" wall. This material might even contain a sampling of the various points-of-view expressed on how planets should be counted for those who feel compelled to do so.

I close with the opinion that a mid-ex style mission to Pluto might resonate much more deeply with the public and with congress if instead of saying "we must complete the reconnaissance of the solar system's planets by sending probes to Pluto" we say "we must BEGIN the reconnaissance of a newly discovered, and hitherto uncharted swath of real-estate in our solar system called the Kuiper belt, of which, Pluto reigns as king.

Respectfully Submitted,
Neil deGrasse Tyson
Department of Astrophysics & Director, Hayden Planetarium
Division of Physical Sciences,
American Museum of Natural History, New York

Appendix F

Resolution of the International Astronomical Union on the Definition of a Planet

IAU Resolution 5A

Adopted 24 August 2006, Prague, Czech Republic

Passed by overwhelming majority of 424 attendees of the session

Definition of a Planet in the Solar System

Contemporary observations are changing our understanding of planetary systems, and it is important that our nomenclature for objects reflect our current understanding. This applies, in particular, to the designation "planets." The word "planet" originally described "wanderers" that were known only as moving lights in the sky. Recent discoveries lead us to create a new definition, which we can make using currently available scientific information.

RESOLUTION 5A (Passed with overwhelming majority vote.)

The IAU therefore resolves that "planets" and other bodies in our Solar System, except satellites, be defined into three distinct categories in the following way:

(1) A "planet" is a celestial body that (a) is in orbit around the Sun, (b) has sufficient mass for its self-gravity to overcome rigid body forces so that it assumes

"YOU'RE FIRED!"

a hydrostatic equilibrium (nearly round) shape, and (c) has cleared the neighbourhood around its orbit.[52]

(2) A "dwarf planet" is a celestial body that (a) is in orbit around the Sun, (b) has sufficient mass for its self-gravity to overcome rigid body forces so that it assumes a hydrostatic equilibrium (nearly round) shape, (c) has not cleared the neighbourhood around its orbit, and (d) is not a satellite.[53]

(3) All other objects except satellites orbiting the Sun shall be referred to collectively as "Small Solar-System Bodies".[54]

52. The eight "planets" are Mercury, Venus, Earth, Mars, Jupiter, Saturn, Uranus, and Neptune.
53. An IAU process will be established to assign borderline objects as dwarf planet or other categories.
54. These currently include most of the solar system asteroids, most trans-Neptunian objects (TNOs), comets, and other small bodies.

Appendix G

New Mexico Legislation Relative to Pluto's Planetary Status

New Mexico 48th Legislature Joint 54 House Memorial
Declaring Pluto a Planet, and March 13, 2007 "Pluto Planet Day"

Introduced by Representative Joni Marie Gutierrez (Democrat, District 33, Dona Ana County), March 8, 2007

WHEREAS, the state of New Mexico is a global center for astronomy, astrophysics and planetary science; and

WHEREAS, New Mexico is home to world class astronomical observing facilities, such as the Apache Point observatory, the very large array, the Magdalena Ridge observatory and the National Solar Observatory; and

WHEREAS, Apache Point observatory, operated by New Mexico state university, houses the astrophysical research consortium's three-and-one-half meter telescope, as well as the unique two-and-one-half meter diameter Sloan digital sky survey telescope; and

WHEREAS, New Mexico state university has the state's only independent, doctorate-granting astronomy department; and

WHEREAS, New Mexico state university and Dona Ana county were the longtime home of Clyde Tombaugh, discoverer of Pluto; and

WHEREAS, Pluto has been recognized as a planet for seventy-five years; and

WHEREAS, Pluto's average orbit is three billion six hundred ninety-five million nine hundred fifty thousand miles from the sun, and its diameter is approximately one thousand four hundred twenty-one miles; and

WHEREAS, Pluto has three moons known as Charon, Nix and Hydra; and

WHEREAS, a spacecraft called *New Horizons* was launched in January 2006 to explore Pluto in the year 2015;

NOW, THEREFORE, BE IT RESOLVED BY THE LEGISLATURE OF THE STATE OF NEW MEXICO that, as Pluto passes overhead through New Mexico's excellent night skies, it be declared a planet and that March 13, 2007 be declared "Pluto Planet Day" at the legislature.

Appendix H
California Legislation Relative to Pluto's Planetary Status

California Assembly Bill HR36 Relative to Pluto's Planetary Status
Introduced by Assembly Members Keith Richman, M.D. (Republican, District 38—northwest Los Angeles County) and Joseph Canciamilla (Democrat, District 11—Contra Costa County, San Francisco Bay Area), August 24, 2006

WHEREAS, Recent astronomical discoveries, including Pluto's oblong orbit and the sighting of a slightly larger Kuiper Belt object, have led astronomers to question the planetary status of Pluto; and

WHEREAS, The mean-spirited International Astronomical Union decided on August 24, 2006, to disrespect Pluto by stripping Pluto of its planetary status and reclassifying it as a lowly dwarf planet; and

WHEREAS, Pluto was discovered in 1930 by an American, Clyde Tombaugh, at the Lowell Observatory in Arizona, and this discovery resulted in millions of Californians being taught that Pluto was the ninth planet in the solar system; and

WHEREAS, Pluto, named after the Roman God of the underworld and affectionately sharing the name of California's most famous animated dog, has a special connection to California history and culture; and

WHEREAS, Downgrading Pluto's status will cause psychological harm to some Californians who question their place in the universe and worry about the instability of universal constants; and

WHEREAS, The deletion of Pluto as a planet renders millions of text books, museum displays, and children's refrigerator art projects obsolete, and represents a substantial unfunded mandate that must be paid by dwindling Proposition 98 education funds, thereby harming California's children and widening its budget deficits; and

WHEREAS, The deletion of Pluto as a planet is a hasty, ill-considered scientific heresy similar to questioning the Copernican theory, drawing maps of a round world, and proving the existence of the time and space continuum; and

WHEREAS, The downgrading of Pluto reduces the number of planets available for legislative leaders to hide redistricting legislation and other inconvenient political reform measures; and

WHEREAS, The California Legislature, in the closing days of the 2005–06 session, has been considering few matters important to the future of California, and the status of Pluto takes precedence and is worthy of this body's immediate attention; now, therefore, be it

Resolved by the Assembly of the State of California, That the Assembly hereby condemns the International Astronomical Union's decision to strip Pluto of its planetary status for its tremendous impact on the people of California and the state's long term fiscal health; and be it further

Resolved, That the Assembly Clerk shall send a copy of the resolution to the International Astronomical Union and to any Californian who, believing that his or her legislator is addressing the problems that threaten the future of the Golden State, requests a copy of the resolution.

Bibliography

Selected Books on Pluto and the Outer Solar System

General Audiences

Asimov, Isaac. *How Did We Find Out About Pluto?* New York: Walker & Company, 1991.

Davies, John. *Beyond Pluto.* New York: Cambridge University Press, 2001.

Elkins-Tanton, Linda. *Uranus, Neptune, Pluto, and the Outer Solar System.* New York: Chelsea House Productions, 2006.

Jones, Tom, and Ellen Stofan. *Planetology: Unlocking the Secrets of the Solar System.* Washington, DC: National Geographic, 2008.

Lemonick, Michael. *The Georgian Star: How William and Caroline Herschel Revolutionized Our Understanding of the Cosmos.* New York: Atlas/Norton, 2008.

Minard, Anne, and Carolyn Shoemaker. *Pluto and Beyond: A Story of Discovery, Adversity, and Ongoing Exploration.* Flagstaff, AZ: Northland Publishing, 2007.

Sparrow, Giles. *The Solar System: Exploring the Planets and Their Moons, from Mercury to Pluto and Beyond.* San Diego: Thunder Bay Press, 2006.

Stern, Alan, and Jacqueline Mitton. *Pluto and Charon: Ice Worlds on the Ragged Edge of the Solar System.* New York: Wiley-VCH, 2005.

Sutherland, Paul. *Where Did Pluto Go?* Pleasantville, NY: Reader's Digest, 2009.

Tocci, Salvatore. *A Look at Pluto.* London: Franklin Watts, 2003.

Weintraub, David A. *Is Pluto a Planet?: A Historical Journey Through the Solar System*. Princeton, NJ: Princeton University Press, 2007.

Children's Books

Asimov, Isaac, Frank Reddy, and Greg Walz-Chojnacki. *A Double Planet?: Pluto and Charon*. Milwaukee: Gareth Stevens Publishing, 1996.

Cole, Joanna. *The Magic School Bus Lost in the Solar System*. New York: Scholastic Press, 1992.

Cole, Michael. *Pluto: The Ninth Planet*. Berkeley Heights, NJ: Enslow Publishers, 2002.

Kortenkamp, Stephen J. *Why Isn't Pluto a Planet?: A Book About Planets*. New York: First Facts Books, 2007.

Orme, David, and Helen Orme. *Let's Explore Pluto and Beyond (Space Launch!)*. Milwaukee: Gareth Stevens Publishing, 2007.

Simon, Tony. *The Search for Planet* X. New York: Basic Books, 1962.

Wetterer, Margaret. *Clyde Tombaugh and the Search for Planet* X. Minneapolis, MN: Carolrhoda Books, 1996.

Sources of Data

Jet Propulsion Laboratories: jttp://ww.jpl.nasa.gov

Lang, Kenneth R. *Astrophysical Formulae*, vols. 1 and 2. New York: Springer-Verlag, 1999.

NASA: http://www.NASA.gov

US Naval Observatory: http://aa.usno.navy.mil/data/

Acknowledgments

The Pluto Files was seven years in the making. Over this time, Pluto became a topic in all conceivable media: television, radio, news articles, comics, op-eds, letters to the editor, and Internet blogs. In sleuthing and selecting the best of these contributions, I am grateful to my research assistant Alison Snyder, whose efforts easily halved the production time this book would have otherwise required. Alison further tracked down and secured permissions from all media sources herein represented, but especially from the letter writers themselves, many of whom were in elementary school when they first wrote, but are now in high school or college.

And so I take this opportunity to thank all of those people—students, teachers, parents, other grown-ups, and colleagues who agreed to lend their correspondence to this volume. *The Pluto Files* exists because of that generosity.

I further thank my brother-in-law Richard Vosburgh, whose expertise on Disney is surely without equal in the land. His research and general base of knowledge greatly enriched my discussions of Pluto the dog and everything else that was Disney in *The Pluto Files*.

Even though he and I stood on opposite sides of the Pluto debate, I benefited from my long friendship, beginning in graduate school, with MIT professor of planetary sciences Richard Binzel. He reliably served and continues to serve as my link to the affairs of the solar system and its motley crew of orbiting objects.

I am further grateful for comments on the manuscript offered by my colleagues Steven Soter at the American Museum of Natural History, Ed Jenkins, of Princeton's Department of Astrophysics, and NASA grammarian Stephanie Schierholz Fibbs.

Parts of *The Pluto Files* were freely adapted from my essays for *Natural History* magazine "The Rise and Fall of Planet X" (June 2003), "Pluto's Honor" (February 1999), and "On Being Round" (March 1997), and from "Requiem for a Solar System," written for *Discover* magazine (November 2006).

Credits

Page x: Copyright © Tribune Media Services, Inc. All rights reserved. Reprinted with permission. Page 2: Copyright © Tribune Media Services, Inc. All rights reserved. Reprinted with permission. Page 5 (top): Courtesy of the Adler Planetarium and Astronomy Museum, Chicago, Illinois. Page 5 (bottom): Timre Surrey Photography, 2007. Page 6 (top): Neil deGrasse Tyson, 2002. Page 6 (bottom): Neil deGrasse Tyson, 2002. Page 8: Public domain. Page 10: Venetia Phair Burney. The author has tried but failed to locate the copyright owner of the photograph of Venetia Burney, and will pay a sensible fee if such person comes forward and proves ownership. Page 11: Royal Astonomical Society / Photo Researchers, Inc. Page 16 (top): Bill Day, 2006, *The Commercial Appeal*. Page 16 (bottom): Copyright © Tribune Media Services, Inc. All rights reserved. Reprinted with permission. Page 19: Paul McGehee, 1986. Page 20: Gary Brookins, 2006, *Richmond Times-Dispatch*. Page 24: Lowell Observatory Archives. Page 26: Lowell Observatory Archives. Page 27: A. J. Dressler and C. T. Russell, "The Pending Disappearance of Pluto," *EOS* 61, no. 44 (1980): p. 690. Copyright © 1980 American Geophysical Union. Reproduced/modified by permission of American Geophysical Union. Page 30: Steven Soter, private communication. Page 32: FoxTrot, Copyright © 2006 by Bill Amend. Reprinted with permission of Universal Press Syndicate. All rights reserved. Page 35 (top): Alison Snyder. Page 35 (bottom): Alison Snyder. Page 36: United States Naval Observatory. Page 37: United States Naval Observatory. Page 38: Vincenzo Zappalá, full astronomer, Astronomical Observatory of Torino, Italy. Page 40: NASA. Page 42 (top): Courtesy of Richard Binzel. Page 42 (below): NASA/JHU/APL/SwRI; image Neil deGrasse Tyson, 2006. Page 43 (top): Neil deGrasse Tyson, 2006. Page 43 (bottom): Neil deGrasse Tyson, 2006. Page 45: Neil deGrasse Tyson, 2006. Page 46: NASA. Page 47: NASA, ESA, H. Weaver (JHU/APL), A. Stern (SwRI), and the HST Pluto Companion Search Team. Page 48: Copyright © 2006 by Jimmy Margulies, *The Record*, and PoliticalCartoons.com. Page 52: NASA. Page 54: David Jewitt and Jane Luu, 1992. Page 56: J. Kelly Beatty, 1996. Page 60: Tom Briscoe, *Small World*. Page 63: Neil deGrasse Tyson, 2000. Page 79 (top): Neil deGrasse Tyson, 2000. Page 79 (bottom): Neil deGrasse Tyson, 2000. Page 86: Marilyn K. Yee / *New York Times* / Redux. Page 91: NASA/ESA/A. Field (STScI). Page 94: *The Joy of Tech* by Nitrozac and Snaggy; www.joyoftech.com. Page 114. Copyright © 2006 by Pat Bagley, *Salt Lake Tribune*, and PoliticalCartoons.com. Page 116: The International Astronomical Union. Page 119: The International Astronomical Union / Lars Holm Nielsen. Page 130: Charles Almon. Page 135: Copyright © 2006 by Bob Englehart, *Hartford Courant*, and PoliticalCartoons.com. Page 140: Courtesy of the California Institute of Technology. Page 150: Copyright © 2006 by R. J. Matson, *St. Louis Post-Dispatch*, and PoliticalCartoons.com. Page 156: Neil deGrasse Tyson, 2007. Page 159: Copyright © 2006 by Aislin, *Montreal Gazette*, and PoliticalCartoons.com. Page 176: Copyright © 2006 by R. J. Matson, *St. Louis Post-Dispatch*, and PoliticalCartoons.com.

Thanks also to all those song writers and the many letter and e-mail writers who gave permission to quote or reprint from their correspondence with me and with others: Mike A'Hearn, Brooke Abrams, Howard Brenner, Don Brownlee, Dan Burns, Siddiq Canty, CCNet, Jonathan Coulton, Timothy Ferris, Will Galmot, John Glidden, Lindsey Greene, Dave Herald, Wes Huntress, Diane Kline, Christine Lavin, Steve Leece, Geoff Marcy, Jeff Mondak and Alex Stangl, Michael Narlock, Bill Nye, Benny Peiser, Robert L. Staehle, Alan Stern, Ian Stocks, Mark Sykes, Madeline Trost, Taylor Williams, and Emerson York.

Index

Page numbers in *italics* refer to illustrations.